MAGNIFICENT UNIVERSE

ALSO BY KEN CROSWELL

The Alchemy of the Heavens (1995)

Planet Quest (1997)

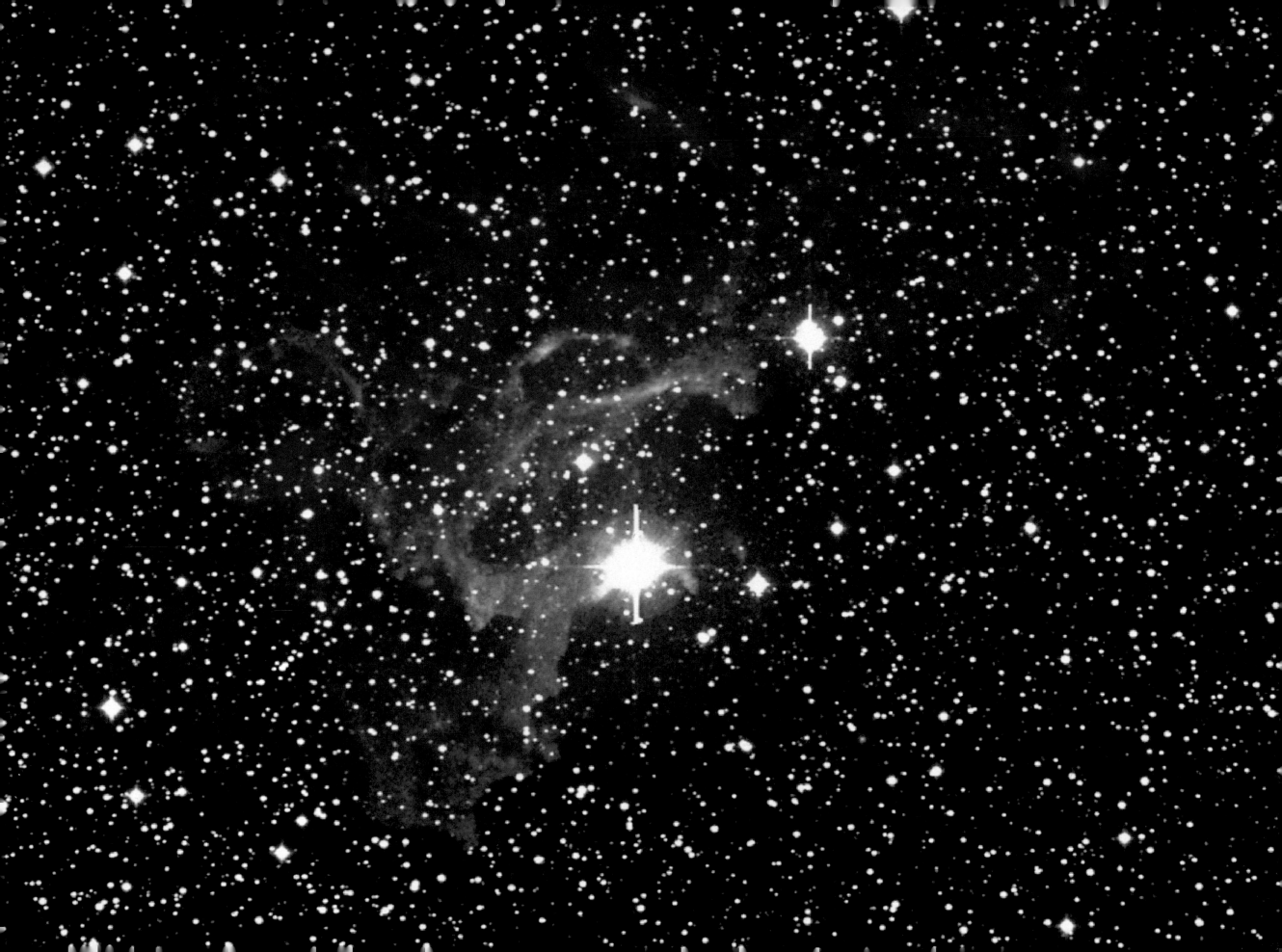

Ken Croswell

SIMON & SCHUSTER

MAGNIFICENT UNIVERSE

SIMON & SCHUSTER
Rockefeller Center
1230 Avenue of the Americas
New York, NY 10020

Copyright © 1999 by Ken Croswell
All rights reserved,
including the right of reproduction
in whole or in part in any form.

SIMON & SCHUSTER and colophon are registered
trademarks of Simon & Schuster, Inc.

Designed by Vertigo Design, NYC

Printed and bound in Great Britain by
Butler & Tanner, Limited

10 9 8 7 6 5 4 3 2

Library of Congress Cataloging-in-Publication Data
Croswell, Ken.
    Magnificent Universe / Ken Croswell.
        p.   cm.
    Includes bibliographical references and index.
    1. Cosmology.   I. Title
QB981.C88 1999
523.1—dc21   99-23875
                                                          CIP

ISBN 0-684-84594-6

CONTENTS

MAGNIFICENT UNIVERSE

Welcome to the universe. *Magnificent Universe* is your ticket to the cosmos, a portrait of the beauty and splendor of the heavens. With it you will journey from the yellow-white clouds of Venus and the orange deserts of Mars to the white rings of Saturn and the icy plains of distant Pluto; from the magenta nebulosity sheathing newborn stars in nurseries like the Orion Nebula to the rainbow-colored remnants cast off by dying stars; from beautiful galactic pinwheels like the Andromeda and Whirlpool Galaxies to the fireworks that ignite when two giant galaxies smash into each other. This voyage will take you across some 12 billion light-years of space, to the very edge of the observable universe.

Before setting off, let's establish where we are. Our planet, the Earth, circles the Sun, from which it receives the light and warmth that keep it alive. Since its birth, the Earth has gone around the Sun 4.6 billion times. Eight other planets also orbit the Sun, and most have one or more moons revolving around them. The planets Mercury and Venus lie so close to the Sun that they are too hot for life, while Mars, Jupiter, Saturn, Uranus, Neptune, and Pluto reside so far away that they are too cold. Smaller bodies—asteroids, comets, meteoroids—also scamper around the Sun. This is our solar system, which we visit in Chapter 1, "The Planets."

Mighty though it is, the solar system's centerpiece—the Sun—is simply a star, similar to the points of light that twinkle at night, and sunlight is really just very strong starlight. The Sun looks so different from other stars because the Earth huddles so close to it. Other stars lie so far away that in order to express their distances, we resort to a gargantuan unit, the light-year—the distance a beam of light speeds through in a year. Shrink the cosmos so that the Earth lies a mere inch from the Sun, and on the same scale even remote Pluto would reside within just forty inches; but a single light-year is so immense it would span an entire *mile*, and the nearest star system to the Sun, Alpha Centauri, would reside over four miles away, corresponding to a true distance of over four light-years. Indeed, if we traveled to Alpha Centauri, the Sun would look like a bright star; if we traveled a few dozen light-years farther, the Sun would look like a faint star; and if we traveled beyond fifty-five light-years, it would disappear from view. But stars would still shine around us—stars red, orange, yellow, white, and blue; stars large and small; stars young and old; stars with planets that may harbor living beings who see our Sun as just another star in the sky. In Chapter 2, "The Stars," we explore how stars are born, live, and die.

These stars speckle our Galaxy, the Milky Way, a giant spiral housing hundreds of billions of stars—two dozen stars for every person on Earth. The Sun is 27,000 light-years from the Galaxy's center, about halfway between the center and the Milky Way's luminous edge. Just as the Moon orbits the Earth and the Earth orbits the Sun, so the Sun orbits the center of the Milky Way, pulling the Earth and the other planets with it. Live eighty years and you will travel through 360 *billion* miles of space—even if you never leave the town in which you were born.

Beyond the shores of the Milky Way shine other galaxies, at least ten of which revolve around our own. The nearest giant galaxy to ours, Andromeda, lies 2.4 million light-years away. The Milky Way, Andromeda, their satellite galaxies, and a handful of other galaxies populate the Local Group, a collection of over thirty galaxies held together by gravity

The Local Group in turn belongs to the Local Supercluster, a vast assemblage of galaxies that stretch across 100 million light-years of space. Here and elsewhere galaxies sport a delightful variety of shapes—fuzzy ellipticals, stunning spirals, amorphous irregulars—which we meet in Chapter 3, "The Galaxies."

The universe today is bigger than it was yesterday, because it is expanding. By running the clock backward, astronomers can deduce that the original universe was small and hot, having arisen in a huge explosion called the big bang, 10 to 15 billion years ago. In Chapter 4, "The Universe," we examine the entire cosmos—its birth, its present expansion, and its future fate.

★

To illustrate this book, I strove to find the very best images of key celestial objects, and I thank those who went to great lengths to show the wonders of the heavens: Stefano Arcella, Jim Baumgardt, Stefan Binnewies, Jay Brausch, Michael Carroll, John Chumack, Licai Deng, Dennis di Cicco, Luke Dodd, Bill and Sally Fletcher, Akira Fujii, George Greaney, Tom Gregory, Tony and Daphne Hallas, Glendon Howell, Franz Kersche, Peter Ledlie, Kenneth Lum, David Malin, Thomas Montemayor, Gerald Rhemann, Peter Riepe, Bernd Schröter, Mike Sisk, Harald Tomsik, and Carl and Chris Weber. Julie Sherwin and Richard Talcott helped me locate several images. I also thank the Anglo-Australian Observatory, the Beijing Astronomical Observatory, the California Institute of Technology, the Canada-France-Hawaii Telescope Corporation, the Dwingeloo Obscured Galaxy Survey, the European Southern Observatory, the Harvard-Smithsonian Center for Astrophysics Supernova Group, the Infrared Processing and Analysis Center, Kitt Peak National Observatory, Lick Observatory, McDonald Observatory, the National Aeronautics and Space Administration, the National Optical Astronomy Observatories, the National Solar Observatory, the National Space Science Data Center, Palomar Observatory, the Space Telescope Science Institute, Tony Stone Images, and the U.S. Geological Survey. These spacecraft contributed images: Apollo 17, Galileo, Hubble Space Telescope, Infrared Astronomical Satellite, Mariner 10, Mars Pathfinder, Pioneer Venus, Viking, and Voyager.

I thank those who read the manuscript and offered comments: Michael Liu, Robert Mathieu, Philip Plait, and Richard Pogge.

Finally, I thank my agent, Lew Grimes, for his support. I thank my German editor, Hans Bender, for proposing this project and my American editor, Stephen Morrow, for his enthusiasm.

THE PLANETS

ONE BY ONE race around the Sun

One by one they race around the Sun: Mercury, swiftest of all, its surface scarred by countless craters; cloud-covered Venus, suffocated beneath air so hot and thick it would kill any life that tried to arise; Earth, beautiful blue, a world of flowing water, mammoth oceans, and vibrant life; Mars, an orange desert with giant volcanoes, dried-up riverbeds, and poles capped by ice; Jupiter, mightiest of all, whose stormy atmosphere boasts a red whirlwind larger than the entire Earth; Saturn, a giant golden globe encircled by bright white rings; Uranus, a green world so faint and distant it eluded the ancients; Neptune, Uranus's twin, a turquoise giant of storms and high winds; and tiny Pluto, icy cold, patrolling the edge of the Sun's planetary domain. On their ceaseless journey around the Sun, most of these worlds carry others with them: the Earth, for example, anchors the Moon, whose light brightens the night, whose gravity stirs the seas. Dodging the planets and their satellites are thousands of asteroids, most between the orbits of Mars and Jupiter, and trillions of comets, which can unfurl majestic tails if they near the Sun.

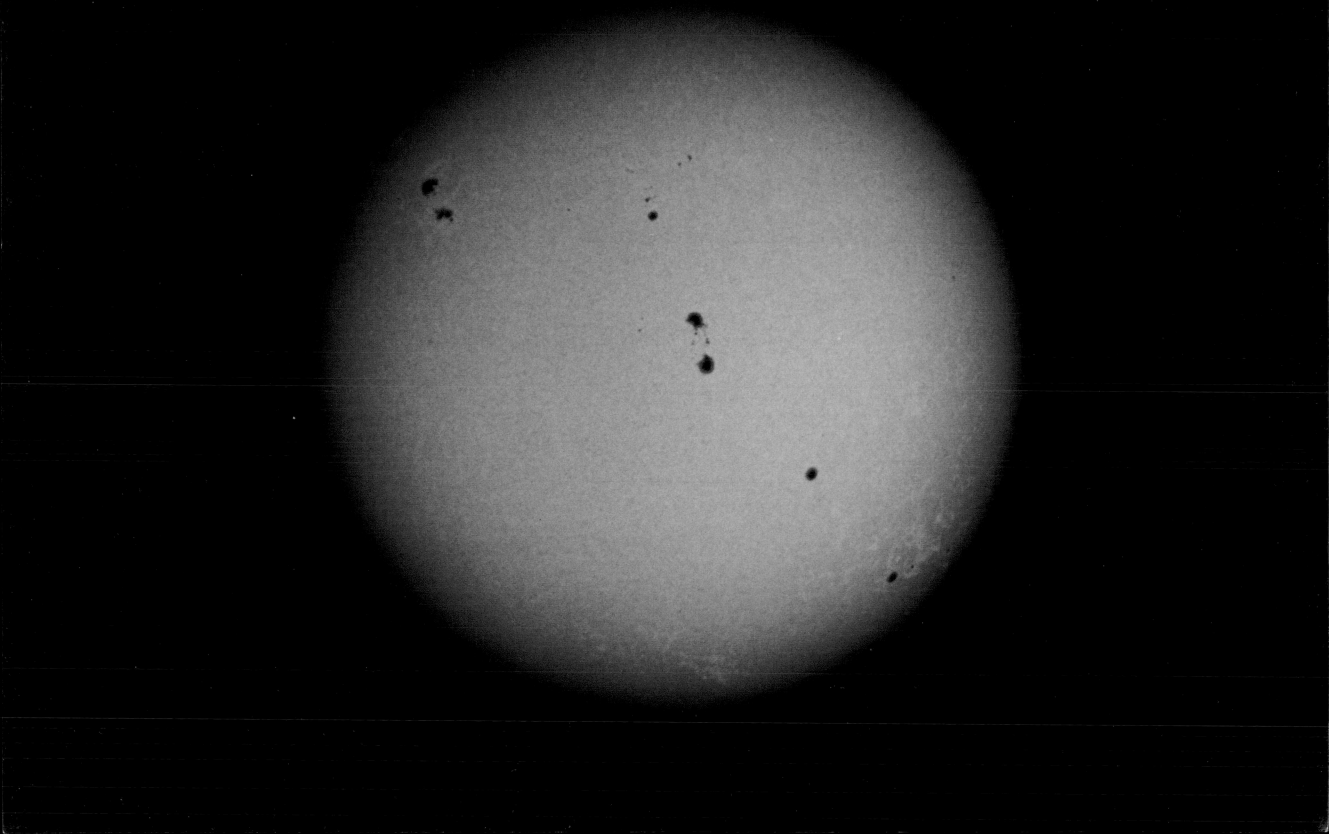

THE SUN

The Sun rules the solar system, hoarding over seven hundred times more mass than all the many worlds that dance around it. The Sun's mass produces the gravity that holds these objects captive, even from as far away as two light-years. The Sun has more than just mass, however; it also emits huge amounts of light that warms and brightens its planets. The Sun generates this energy from nuclear power, fusing nuclei of hydrogen, the lightest element in the universe, into nuclei of helium, the second lightest element. This reaction transforms a tiny amount of mass into pure energy. Each second, the Sun converts 600 million tons of hydrogen into helium. This nuclear reaction, at the Sun's center, releases deadly x-rays and gamma rays. But as they try to escape the Sun's core, the x-rays and gamma rays bounce to and fro against the Sun's own material and lose energy. As a result, a typical x-ray or gamma ray takes 30,000 years to reach the solar surface. By the time the once-lethal radiation finally wins the battle, it is so exhausted that it emerges as the relatively low-energy visible light which changes night into day.

Threading the Sun are magnetic fields that generate sunspots on its surface and flares erupting above it. The number of spots waxes and wanes approximately every eleven years. From 1645 to 1715, however, nearly all spots vanished. During this period, called the Maunder minimum, Earth cooled—probably because the Sun faded. Surrounding the Sun's surface is the corona, hot but tenuous gas extending millions of miles, and blasting through the corona is the solar wind, a stream of charged particles that shoots beyond Pluto.

As a star, the Sun distinguishes itself from the planets by generating its own light, whereas the planets merely reflect the light from the Sun. Nevertheless, the Sun and its planets formed at the same time, 4.6 billion years ago. Somewhere in our Galaxy, a cloud of gas and dust collapsed under the weight of its gravity. Much of this material fell into what was to become the Sun, but some remained in orbit, in a swirling disk. Dust and ice in the disk stuck together and grew into asteroids and comets, which then collided with one another to give birth to the nine planets. These still revolve around the Sun in nearly the same plane and in the same direction, counterclockwise as viewed from above the Earth's north pole.

Different substances condensed out of the gas at different distances from the Sun. The inner part of the disk spun fast, so friction heated it; therefore, in the inner disk, only hardy substances with high melting points—rock and iron—condensed and formed planets. Thus, the solar system's four inner planets—Mercury, Venus, Earth, and Mars—became worlds of rock and iron. They are small because the disk contained little of these materials. In the outer solar system, however, the disk was cool, so ice also condensed, which was far more common than the rock and iron. As a result, four outer planets—Jupiter, Saturn, Uranus, and Neptune—grew into giants, and their great gravity stole some of the hydrogen and helium gas that pervaded the disk, further augmenting their sizes. Scattered about the solar system was debris left over from the planets' formation—asteroids, comets, and Pluto.

MERCURY bears
crater after crater

MERCURY

The Sun's first planet is something of an acquired taste. Mercury drapes itself in a gray surface that bears crater after crater, prompting one critic to call it a world only a confirmed crater counter could love. A poll once asked space enthusiasts to name the worst spacecraft ever launched; one respondent said, "Anything having to do with Mercury."

FACING PAGE: Cratered Mercury boils beneath brutal sunlight.

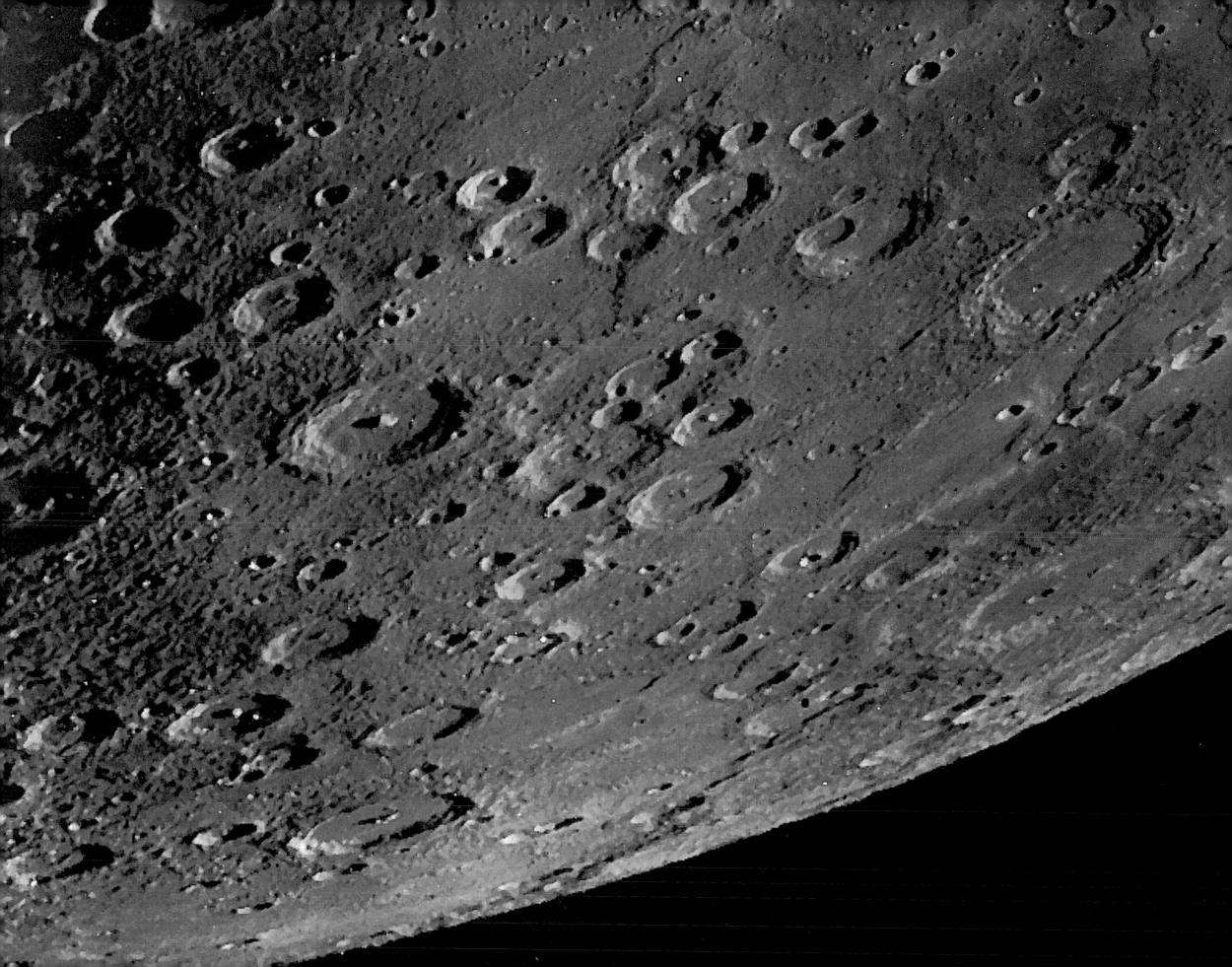

Stymied by Mercury's proximity to the Sun, astronomers learned little. They knew it was small—just over a third the size of Earth—and suspected it was hot. They also knew it orbited the Sun in just 88 days, making Mercury's year a quarter as long as Earth's. Only in 1965, however, did they discover its rotation period, 58.6 days. Before then, they thought it spun every 88 days, the same as its year. This equality between day and year would have meant that one hemisphere forever faced the Sun and fried, while the other forever faced away and froze.

In 1974 and 1975, the Mariner 10 spacecraft flew past Mercury three times and delivered most of what is now known about the planet. Even though Mercury's day does not equal its year, the planet exhibits an enormous temperature range, from 800 degrees Fahrenheit at hottest—sufficient to melt lead—to −300 degrees Fahrenheit at coldest. The extreme temperature range stems from Mercury's long day and absence of an appreciable atmosphere to insulate itself. High temperature, in turn, destroys any hope of atmosphere, since heat imparts high speeds to airborne molecules and causes them to escape the planet's weak gravity. Because of Mercury's heat, scientists in 1991 were astonished to discover water ice at the planet's poles. This ice hibernates in crater floors that never see the Sun.

UNLIKE EVASIVE MERCURY, Venus proudly decorates the changing colors of twilight

VENUS

The Sun's second planet, Venus, at first looks more inviting. It is certainly more prominent. Unlike evasive Mercury, Venus proudly decorates the changing colors of twilight as a brilliant morning or evening star that outshines every other heavenly body but the Sun and Moon. Venus can even cast a shadow. Because of its radiance, ancient people named it for the goddess of love and beauty—and modern people occasionally think it a UFO.

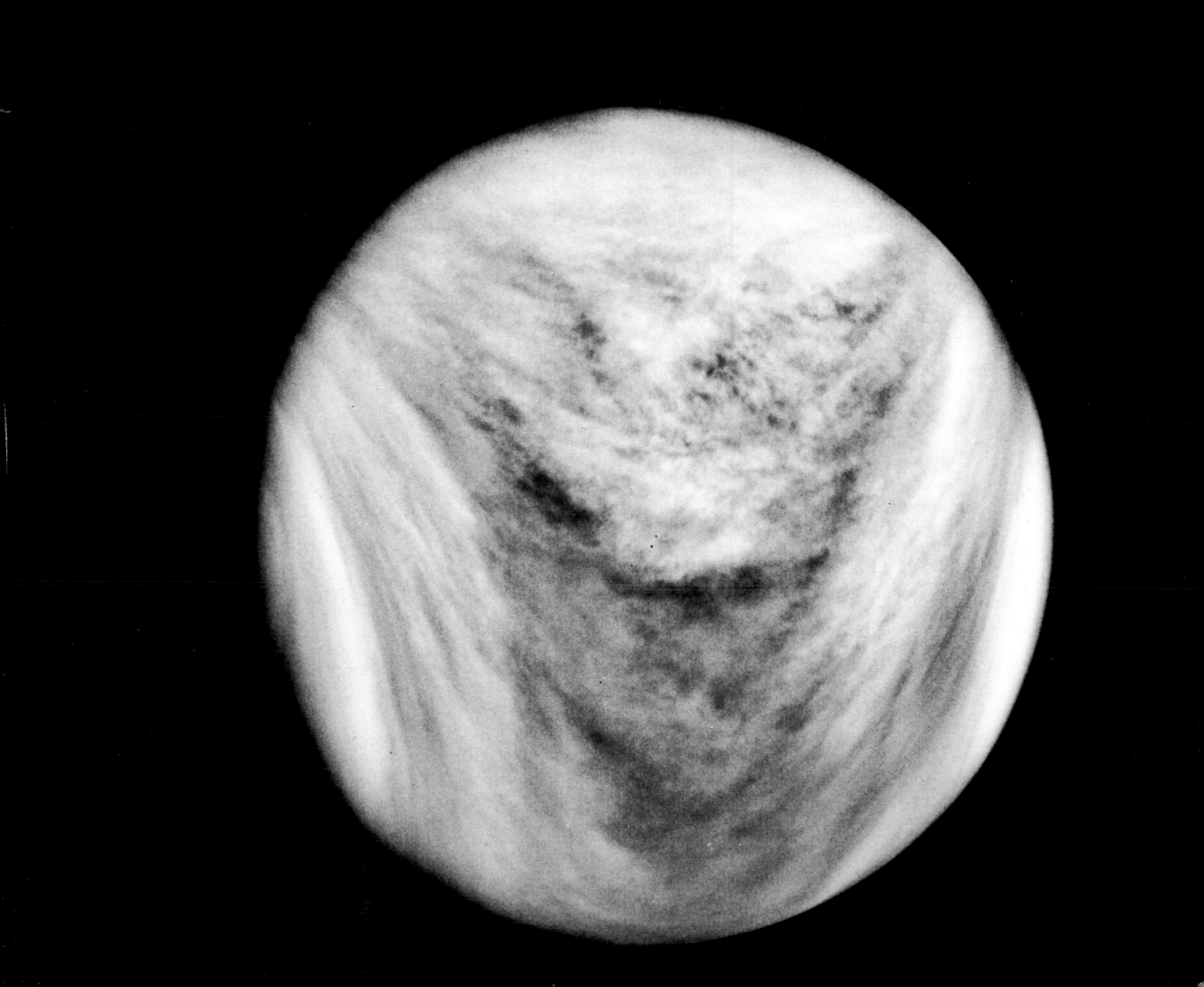

Venus once evoked visions of life, for in some ways the planet so resembles our own that it was called Earth's twin. Venus nearly equals Earth's size, mass, and density. It passes closer to Earth than does any other planet, which is one reason Venus looks so bright. And like Earth it has an atmosphere and clouds. The clouds also contribute to Venus's brightness, for they reflect most of the sunlight striking them.

The clouds cloak Venus's surface, so scientists—and science fiction writers—were free to speculate about what the clouds hid. Clouds suggested water, so some scientists pictured dense tropical jungles, warm and steamy, teeming with creatures large and small, while others thought a vast ocean covered the entire planet, broken only by a few craggy peaks here and there.

The real Venus, though, is a hot, dry desert smothered beneath a hot, thick atmosphere. Because Venus is so close to Earth, it was the first planet to receive a spacecraft, Mariner 2 in 1962. This and subsequent missions painted a portrait of a hellish world. The atmosphere is over ninety times thicker than Earth's; its clouds consist not of water, as on Earth, but of sulfuric acid–battery acid; and its atmosphere is full of carbon dioxide, the same gas we get rid of when we breathe. The carbon dioxide produces a terrible greenhouse effect that makes Venus, not Mercury, the hottest planet in the solar system: sunlight penetrates the atmosphere (if you stood on Venus, it would look like a heavily overcast day on Earth) and heats the surface, but the carbon dioxide prevents the heat from escaping. As a result, the surface of Venus is 860 degrees Fahrenheit.

Venus and Earth actually possess similar quantities of carbon dioxide, but little of Earth's seeps into the air. Instead, it resides in corals and carbonate rocks, like limestone, because life incorporates it into shells, and rainfall washes it from the atmosphere. Venus has neither life nor rain to remove atmospheric carbon dioxide, so whereas carbon dioxide accounts for only 0.035 percent of Earth's atmosphere, it makes up 96.5 percent of Venusian air.

A few spacecraft have plunged through the thick atmosphere and briefly glimpsed Venus's rock... surface. Other spacecraft, such as Magellan, have cir... cled the planet from afar and peered at the surface b... radar. These images reveal a world ruled by lava–volca... noes, lava flows, and lava plains. A seven-mile-high mountain range, Maxwell Montes, rises near the planet's north pole. Venus possesses the longest river channel in the solar system, Baltis Vallis, which stretches over 4,200 miles, greater than the distance between Anchorage and Miami; but it was carved b... lava, not water.

Billions of years ago, Venus may have had water too, perhaps even oceans, which may have given rise to life. At that time, the Sun was fainter and Venus coole... so ancient Venus may have resembled modern Earth. Rain removed carbon dioxide from the atmosphere, keeping Venus mild. As the Sun brightened, though, Venus heated up and its oceans evaporated. Rain no longer fell, so carbon dioxide gas accumulated and the planet grew even hotter. No wonder, then, that scientists worry about another planet whose atmospheric carbon dioxide abundance is increasing—the Earth.

EARTH

Once regarded as the supreme center of the universe around which all else revolved, the Earth is now known to be the Sun's third and most precious planet, a world that speeds around the Sun at 67,000 miles per hour and carries a cargo unique in the solar system: a vibrant web of millions of different living species. Earth is the largest of the four inner planets. It has a large iron core enveloped by a rocky mantle and crust. Currents in the core generate a magnetic field that shelters life from the solar wind. Interaction between the solar wind and the magnetic field creates aurorae, the northern and southern lights.

FACING PAGE: Earth is the third planet from the Sun.

FOLLOWING PAGES:

Page 12. Volcanoes help keep Earth warm enough for life by emitting carbon dioxide, a greenhouse gas.

Page 13. A rich web of life blankets Earth.

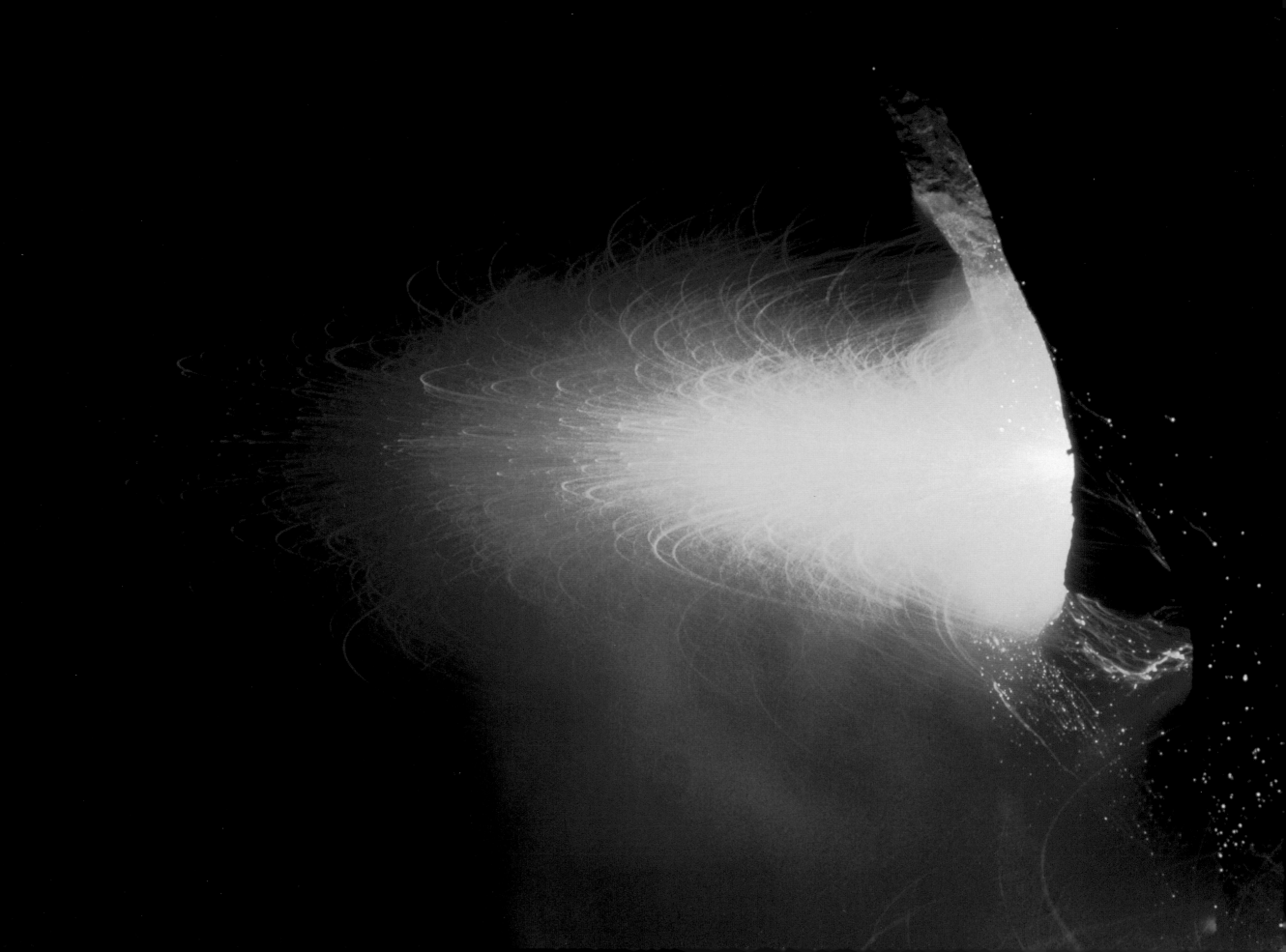

ONCE REGARDED as the
supreme center of the
universe around which
all else revolved, the
Earth is now known to
be the Sun's third and
most precious planet

Earth has a rich supply of liquid water, the substance in which terrestrial life probably first arose. Water blankets 71 percent of the Earth's surface, most in oceans, but some in rivers, lakes, and ice sheets. Surrounding the Earth is a thin atmosphere of nitrogen (78 percent) and oxygen (21 percent). The atmosphere not only powers life but also insulates Earth from wild day-to-night temperature swings; shields Earth from the countless meteoroids that bombard it; maintains the ozone layer that protects land-based life from the Sun's deadly ultraviolet rays; and bears greenhouse gases—especially carbon dioxide and water vapor—that raise the Earth's temperature 60 degrees Fahrenheit, thereby transforming what would be a barren ice-covered planet into a luxurious world brimming with life.

No other planet in the solar system abounds with so much atmospheric oxygen, and billions of years ago neither did Earth itself. At that time, the atmosphere was primarily carbon dioxide, water vapor, and nitrogen, all vented from volcanoes. The carbon dioxide and water vapor kept the Earth warm even though the Sun was then fainter. As the Sun brightened, rainfall removed the carbon dioxide from the atmosphere, so Earth remained mild. Meanwhile, another phenomenon began to alter the air. The first life arose in the oceans some 4 billion years ago, and photosynthesis consumed carbon dioxide and produced oxygen. The oxygen at first combined with oceanic rocks, but about 2 billion years ago, with the rocks saturated, the oxygen bubbled into the atmosphere. Some of the oxygen formed ozone, which guarded the surface from solar ultraviolet radiation and allowed life to advance from sea to land.

THE MOON

The Earth's nearest celestial neighbor may also have helped. No other small planet has such a large satellite. Mercury and Venus are moonless, and the moons of Mars are minuscule. But the Moon is over a quarter the size of Earth, and lunar tides may have pushed life from the oceans onto the land. In the past, the Moon was closer and these tides were stronger. Today the Moon revolves around the Earth faster. The Moon then revolved around the Earth faster. The Moon revolves only once a month; the word "month" derives from "Moon." The Moon rotates as fast as it revolves, so the same side always faces Earth.

The Moon has long been linked to madness, or "lunacy." Even today, people sometimes superstitiously blame the Moon, especially the full Moon. However, most of the time that one hears "There must be a full Moon tonight," the Moon is not actually full.

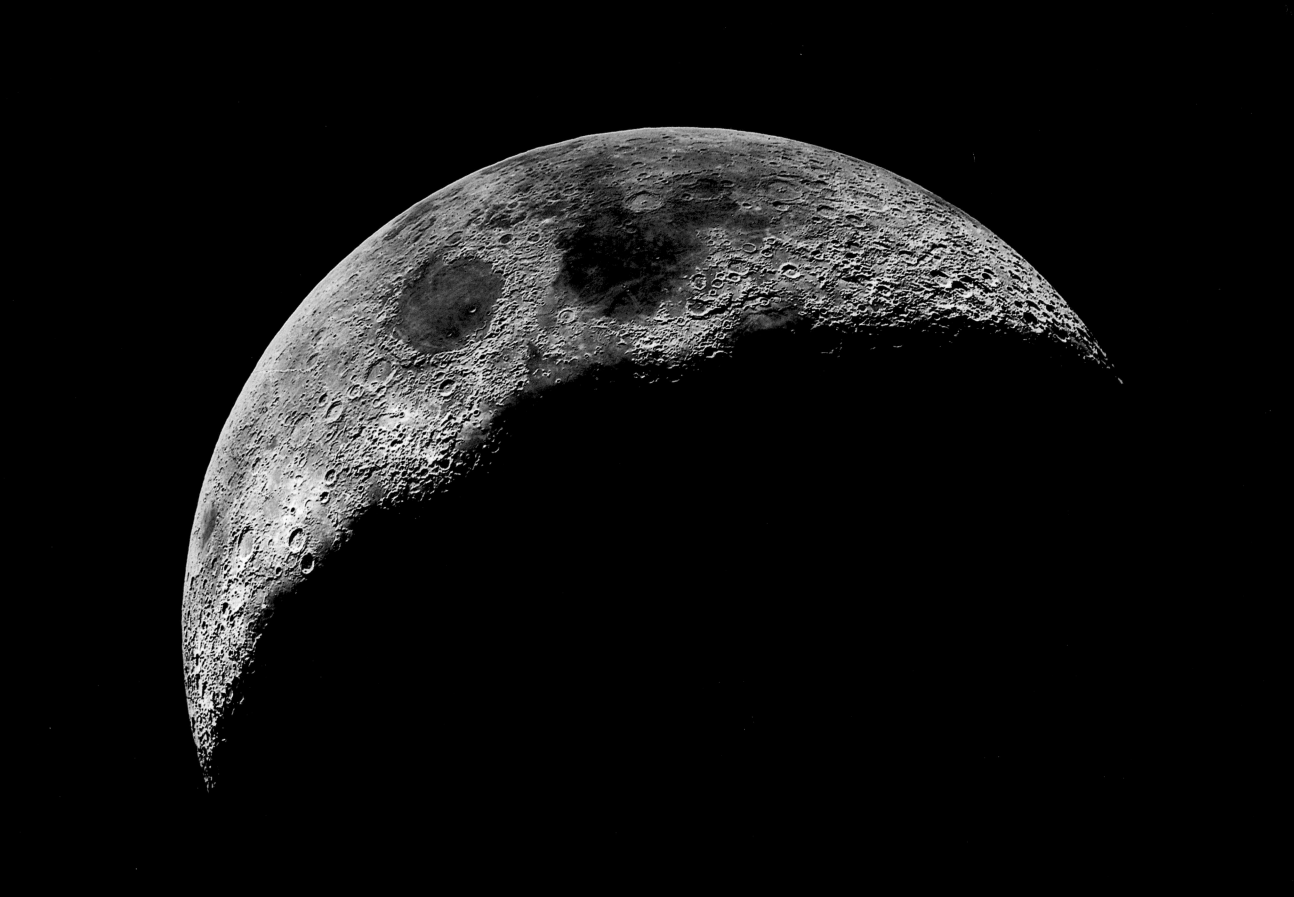

lands, which are ancient, dating back some 4 billion years, and the smoother maria, or lunar seas, which formed somewhat later, when lava overflowed these regions. In contrast, the Moon's far side consists almost exclusively of cratered highland terrain. The Moon boasts the solar system's largest impact basin, the South Pole-Aitken Basin, which is 8 miles deep and 1,600 miles across. Although the rocks gathered by Apollo astronauts are incredibly dry, polar craters, perpetually shielded from sunlight, seem to harbor ice, just as Mercury's poles do.

A BEAUTIFUL crescent suspended in a twilit sky can stir our hearts

Not only does the Moon's cratered face testify to the solar system's violent past, but so does the Moon's very existence. According to the leading theory for the Moon's origin, a Mars-sized object smashed into the young Earth and splattered material from the Earth's mantle into orbit, where it conglomerated into a large satellite. This is why the Moon's composition matches that of the Earth's mantle: much rock, little iron.

MARS

The fourth planet from the Sun looks like a drop of blood, prompting the ancients to name it for the god of war. Mars is indeed ruddy for the same reason blood is. Blood turns red when the iron-bearing compound hemoglobin joins with oxygen, and Mars is reddish because the same two elements join to form iron oxide, or rust, on the Martian deserts.

Mars is best known for its Martians, conjured up by imaginative Earthlings. Foremost among these was wealthy Bostonian Percival Lowell, who in 1894 founded an observatory in Arizona to study the neighbor world. Lowell thought he saw canals the Martians had built to ferry water from the planet's polar caps to its equator. Lowell's visions excited the public but antagonized other astronomers; nevertheless, long after Lowell died, many scientists believed that Mars at least had plants.

In 1965, even these modest hopes were dashed when the Mariner 4 spacecraft flew past the planet and revealed a cratered landscape that looked like the lifeless Moon. This and later missions, including several that landed on Mars, depicted a frozen world with an atmospheric pressure less than 1 percent of that on Earth. There is almost no oxygen to breathe and no ozone to protect the surface from the Sun's ultraviolet rays.

Still, Mars offers far friendlier surface conditions than any other planet but Earth. Furthermore, certain red planet traits uncannily mirror the Earth's. For example, Mars spins once every 24 hours, 37 minutes, and the planet's axis tilts 25.2 degrees, so Martian days and seasons resemble terrestrial ones, except the latter last nearly twice as long, since Mars takes nearly twice as long to orbit the Sun, 687 days.

Mars exhibits a north-south asymmetry: the northern hemisphere is fairly smooth, the southern hemisphere heavily cratered. The two hemispheres recall the two types of lunar terrain, the smooth lunar

IF MARS was once a warm, wet world where life flowered, what went wrong?

...eds, of mana, and the cratered highlands. The north-ern hemisphere must be younger, since lava flows have eradicated most of its craters. Yet the old southern hemisphere, battered and cratered though it is, pre-serves signs of a kinder, gentler Mars: valleys carved by running water, indicating that billions of years ago, Mars was warmer and wetter. During that ancient epoch, Mars may have sprouted life, whose fossils may still exist.

If Mars was once a warm, wet world where life flowered, what went wrong? Mars resides farther from the Sun's heat than does the Earth, which partially explains the frigid Martian climate. When Mars was young, however, its volcanoes spewed carbon dioxide and water vapor, triggering a greenhouse effect that warmed the planet, allowing water to flow. But Mars was doomed. Volcanoes derive their strength from a planet's internal heat, which in turn depends on the planet's size. Although Earth and Mars were both born hot, Mars is only half as big, so its interior cooled faster, just as a small, freshly baked roll cools faster than a large loaf of bread. As the Martian interior cooled, the volcanoes shut down, the atmosphere thinned, the greenhouse effect diminished, and the planet froze, killing any life. Today, Martian air is still mostly carbon dioxide, but so thin that the greenhouse effect lifts the temperature only 10 degrees Fahrenheit.

As if to prove their past power, huge volcanoes still tower above the red plains of Mars. The largest, which dwarf Mount Everest, cluster together in the Tharsis bulge, a blister on the Martian sphere. The formation of Tharsis apparently cracked the surface, for running away from Tharsis is a long canyon, Valles Marineris, which if transported to Earth would stretch from Cincinnati to San Francisco.

Circling the red planet are two moons named for the horses that escorted the chariot of Mars, Phobos ("fear") and Deimos ("terror"). These small, cratered worlds, irregularly shaped, were first seen by American astronomer Asaph Hall in 1877, but Irish author Jonathan Swift had actually written of both a century and a half before, in his 1726 satire *Gulliver's Travels*. Even earlier, German astronomer Johannes Kepler had predicted two moons for Mars, reasoning that since Earth had one and Jupiter was then known to have four, Mars should have a number between, two or three. He chose two, as did Swift, who probably meant to ridicule Kepler's logic.

Both moons lie much closer to Mars than the Moon does to Earth, and they therefore circle the planet much more quickly. Phobos, the larger and nearer, orbits every 7.7 hours, faster than Mars spins, so a Martian colonist would see the satellite rise not in the east but in the west. More distant Deimos takes 30 hours to revolve. Both moons are dark, like most aster-oids, suggesting that Mars snatched them from the neighboring asteroid belt.

FACING PAGE: Mars once sparked the hope of a living planet near Earth.

FOLLOWING PAGES:

Pages 20-21. Now cold and dry, Mars may have once been warm, wet, and alive.

Page 22. Potato-shaped Phobos circles Mars.

Page 23. A small moon tars beside asteroid Ida.

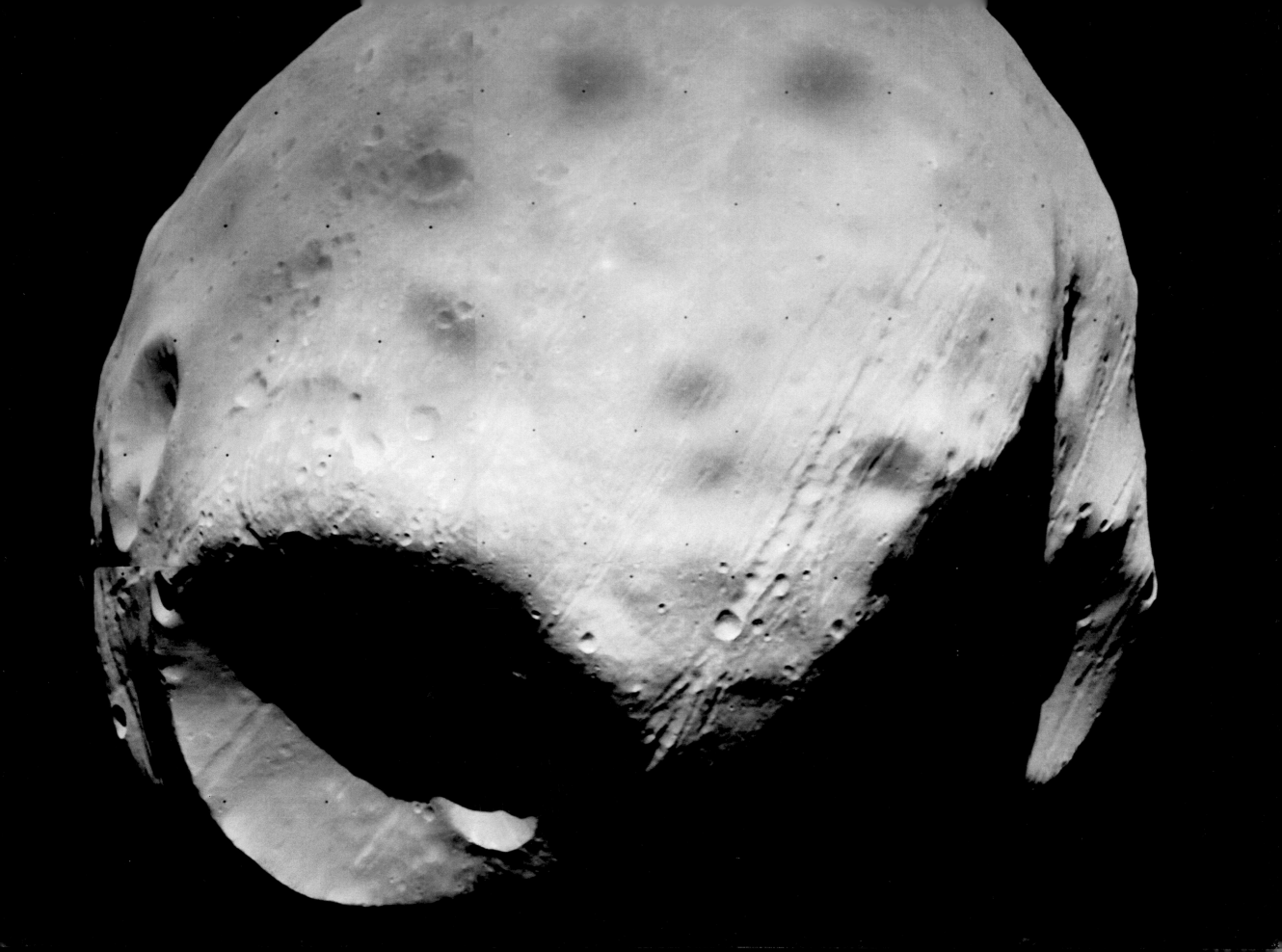

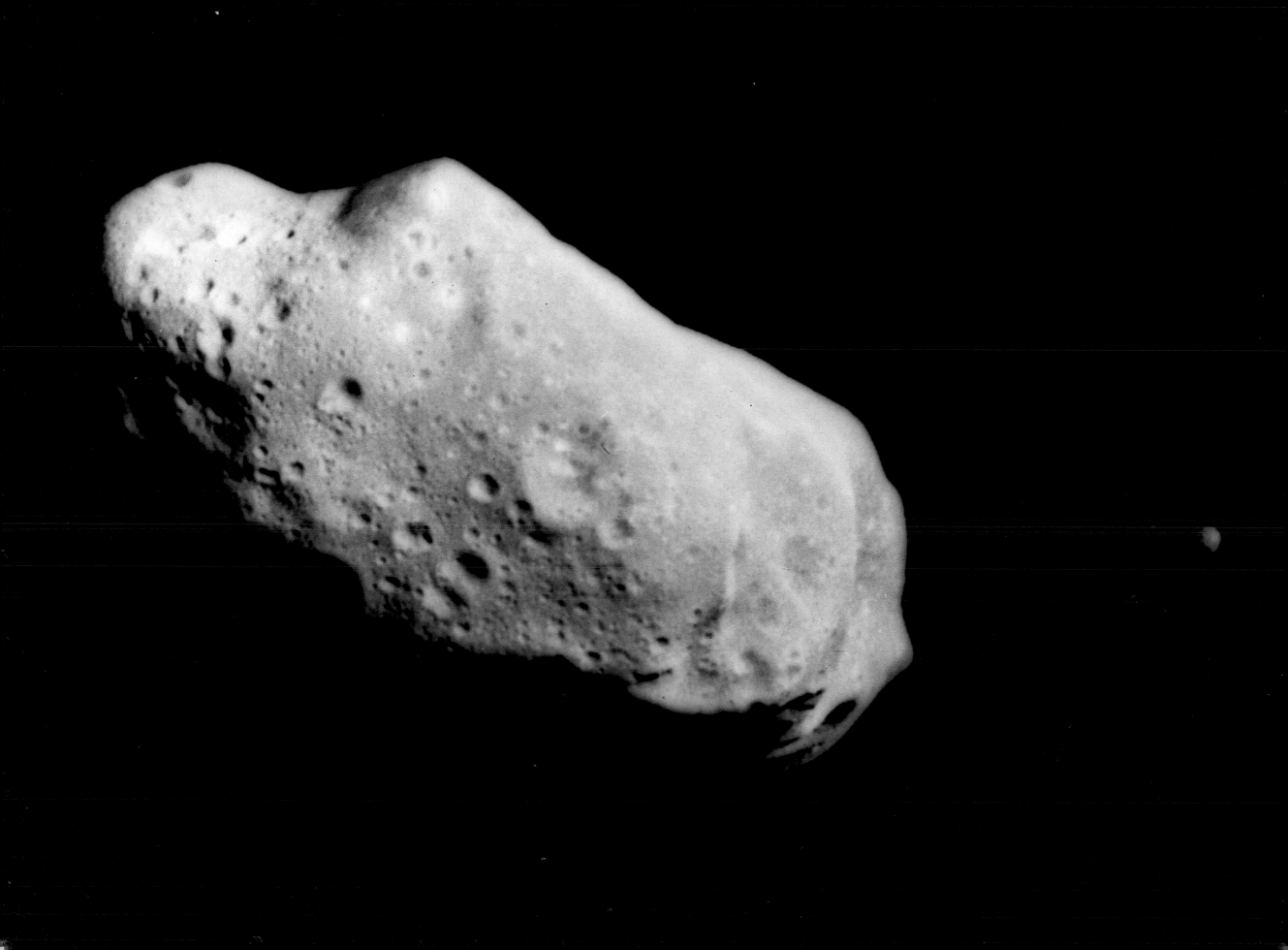

Asteroids are small, rocky bodies. The first and largest, six-hundred-mile-wide Ceres, was discovered in 1801 and thought to be a planet. But then a second asteroid, Pallas, was found in 1802, followed by a third and fourth, Juno and Vesta, in 1804 and 1807. Today thousands have been catalogued, most lying between the orbits of Mars and Jupiter. They are victims of Jupiter's immense gravity. When the Sun was young, still surrounded by a disk of gas and dust, Jupiter's gravity prevented the material inside its orbit from aggregating into a full-fledged planet.

Although science-fiction movies sometimes portray asteroid belts choked with deadly debris, asteroids are actually so far apart that a million spacecraft could fly blind through the asteroid belt and likely not suffer a single collision. Nevertheless, in 1972, when the first spacecraft trespassed into the realm of the asteroids, scientists feared the craft would smash into smaller material, dust and pebbles, too faint for astronomers to see. Fortunately, this zone turned out to be fairly clean, and several spacecraft have successfully traversed the asteroid belt and reached the outer planets.

In 1991 and 1993, one such spacecraft, Jupiter-bound Galileo, photographed asteroids Gaspra and Ida, and in 1997, the Near Earth Asteroid Rendezvous (NEAR) spacecraft sailed past Mathilde. These asteroids proved heavily cratered and irregularly shaped, like the moons of Mars. This is to be expected, since asteroids get hit by other asteroids and most have too little gravity to pull themselves into spheres. Ida even has its own moon, a tiny world just a mile across. It is the smallest moon ever found.

Not all asteroids stay within the asteroid belt. Some venture closer to the Sun and even cross the orbit of Earth. Human beings owe their existence to one such asteroid that collided with Earth 65 million years ago and killed off the dinosaurs, thereby letting new forms of life take their place. But what gave life

can also take it away: a similar impact today might wipe out civilization. Other asteroids, called Trojans, lie on the other side of the asteroid belt from Earth, at Jupiter's distance from the Sun, 60 degrees ahead of and behind the planet.

JUPITER

The ancients named Jupiter well. Because of the planet's brilliance (among planets, normally second only to Venus) and its slow, majestic movement (it takes twelve years to orbit the Sun), the planet suggested the king of the gods. The ancients knew nothing of Jupiter's gargantuan size, but it harbors over twice the mass of all the other planets combined, 318 Earth masses. The king of the gods is also the king of the planets.

THE KING of the gods is also the king of the planets

Jupiter stands apart from the inner planets not only because of its great size but also because of its unearthly composition. Whereas the Earth and the other inner planets are dense rock-iron worlds, Jupiter consists of more ethereal material. The planet is a gas giant, composed primarily of hydrogen and helium, the lightest and most common elements in the universe. On Earth they are gases, but Jupiter's enormous weight squeezes most of the hydrogen into a metal that could conduct electricity. The hydrogen and helium envelop a rock-water core. Though this core has roughly ten times the mass of the Earth, it holds only a few percent of Jupiter's total mass.

FACING PAGE: Jupiter outweighs all of the Sun's other planets put together.

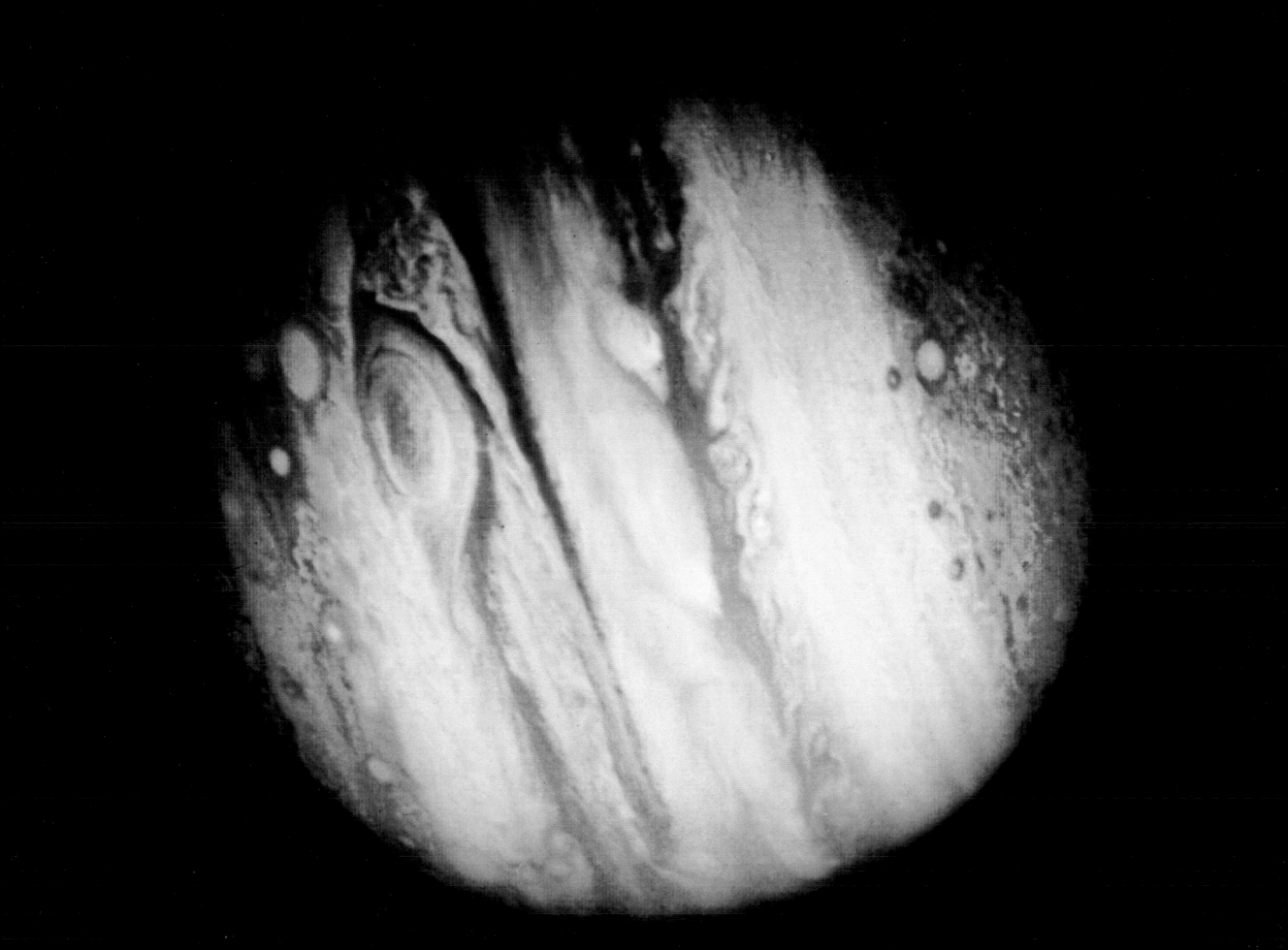

Jupiter radiates more energy than it receives from the Sun. Some of Jupiter's excess energy is left over from its birth, and some arises because its helium is sinking and giving off energy, like water falling over a dam. This energy helps keep Jupiter's atmosphere stormy, turbulent, and colorful. Furthermore, Jupiter spins faster than any other planet, once every 9 hours and 55 minutes, which also stirs up the atmosphere. The planet's banded disk sports a red spot bigger than Earth that has persisted for centuries. Because of Jupiter's rapid rotation and its fluffy hydrogen-helium composition, the planet is flattened, its equatorial diameter exceeding its polar diameter by thousands of miles.

Although Jupiter is lifeless, in 1992 George Wetherill of the Carnegie Institution of Washington suggested that both it and its neighbor Saturn may be responsible for intelligent life on Earth, because the mighty gravity of Jupiter and Saturn long ago ejected trillions of comets out of the solar system—leaving relatively few of these deadly objects to strike the Earth, and sufficient time between major impacts for intelligence to evolve. Two years later, as if to confirm this idea, Jupiter took a direct hit from Comet Shoemaker-Levy 9, which left the guardian world scarred for months.

Jupiter rules a miniature solar system, a retinue of at least sixteen satellites. The four largest, which Galileo Galilei spotted in 1610, gave him ammunition to shoot down the Catholic Church's assertion that all worlds circled Earth. From innermost to outermost, the four Galilean satellites are Io, Europa, Ganymede, and Callisto (mnemonic: I Eat Graham Crackers). Ganymede is larger than either Mercury or Pluto; in fact, it is the largest moon in the solar system.

Of the four Galilean satellites, Io is the most fiery. In 1979, the Voyager 1 spacecraft discovered nine erupting volcanoes on the moon, eight of which were still going strong when Voyager 2 flew by four months later. The volcanoes spew so prolifically that their lava quickly erases craters. Io owes its volcanic strength to

Jupiter, whose tides squeeze, stretch, and heat the moon's interior.

Europa, the smallest Galilean satellite, holds not fire but ice. Beneath its icy surface may lurk an ocean of liquid water. If the ocean floor bears heat vents powered by Jovian tides, those vents could provide energy for life, which would blossom not in the light but in the dark. On Earth heat vents beneath the ocean nourish creatures that never see the Sun. Still, life in Europa's hypothetical ocean remains speculative and may prove just as chimerical as the beings once envisioned on Venus and Mars.

Twelve other moons circle Jupiter, four inside the Galilean orbits, eight outside. One of the inner moons, egg-shaped Amalthea, orbits close to Io and is colored reddish by its volcanoes. The outer moons may be captured asteroids or comets. The four most distant of these all revolve backward, suggesting that they indeed did not form with Jupiter. In addition to moons, Jupiter possesses a ring, but one so faint that it hardly rivals those of Saturn.

JUPITER RULES a miniature solar system, a retinue of at least sixteen satellites

FACING PAGE: Jupiter's moon Io sports active volcanoes.

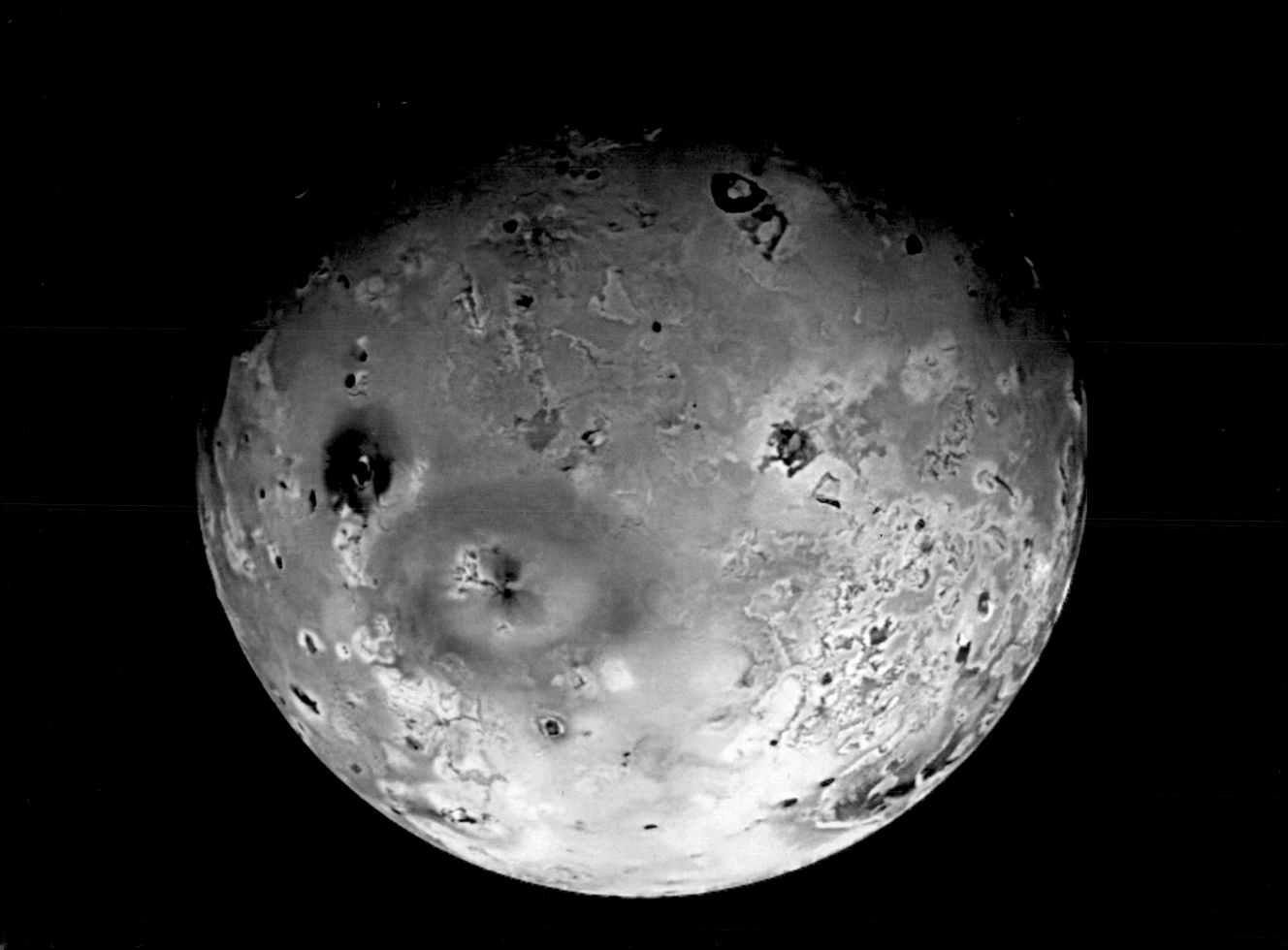

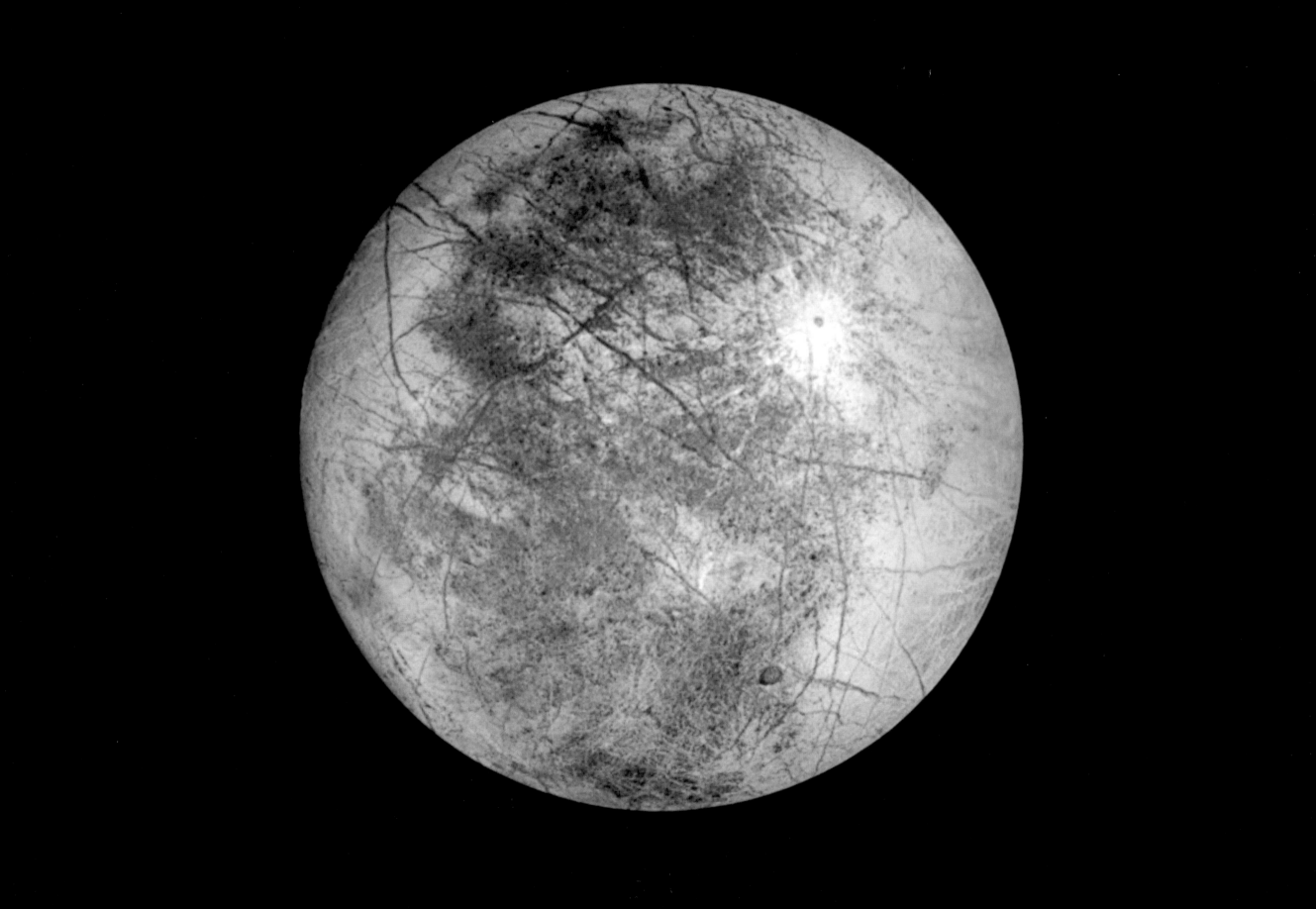

SATURN SEEMS the most cheerful of planets, thanks to its bright, resplendent rings

Saturn once marked the solar system's frontier, the farthest planet astronomers knew. It takes 29½ years to complete a cycle, longer than any other classical planet. Though named for the somber god of time, Saturn seems the most cheerful of planets, thanks to its bright, resplendent rings, discovered in the 1600s.

As the solar system's other gas giant, Saturn resembles a smaller Jupiter. It is mostly hydrogen and helium surrounding a rock-water core. Whereas Jupiter has 318 Earth masses, Saturn has 95. Because of its lower mass and weaker gravity, Saturn cannot compress its hydrogen and helium as much as Jupiter does, and thus Saturn's density is so low that if the planet were placed in an ocean large enough, it would float. And like Jupiter, Saturn emits more energy than the Sun provides, perhaps because helium is separating from the lighter hydrogen and falling toward Saturn's core, thereby releasing energy and warmth.

Saturn's golden globe occasionally displays white spots, but the planet lacks the stormy turbulence that marks its big brother. Saturn more than compensates by adorning itself with stunning rings. From one side of Saturn to the other, the three brightest rings span 170,000 miles—nearly three-fourths of the distance between the Moon and Earth. The rings consist of countless ice particles. They may be the wreckage of a moon or comet that strayed too close to Saturn and got torn apart by its gravity, which pulled harder on the near side of the

luckless object than on the far side. Or the rings may be the remains of a small moon that got smashed to bits by a comet. The rings are so thin that they disappear from terrestrial view when the planet turns exactly equator-on to Earth. These ring-plane crossings occur twice during Saturn's year, or about once every fifteen Earth years. The next happens in 2009.

During these ring-plane crossings, astronomers can better see Saturn's surroundings and glimpse its moons, most of which were first spotted at such times.

The moons number at least eighteen, more than revolve around Jupiter. The king of the Saturnian moons is the appropriately named Titan, a world larger than Mercury and Pluto and the second largest moon in the solar system, after Jupiter's Ganymede. Titan first made its unusual nature known in 1944, when American astronomer Gerard Kuiper reported methane gas around the satellite. Although it seemed thin, the atmosphere was the first ever found around a mere moon. In 1980, Voyager 1 discovered that Titan's true atmosphere is actually thicker than Earth's. Furthermore, it consists mostly of the predominant gas surrounding Earth, nitrogen. The previously known methane exists, too, but constitutes only a few percent of Titan's air. Unfortunately, the atmosphere also contains orange haze that blocked Voyager's view of the surface, but in 2004 the Saturn-bound Cassini spacecraft will send a probe to sail through the atmosphere and possibly land on the surface.

Titan is especially intriguing because in some ways it resembles ancient Earth: it has nitrogen and organic compounds, but no atmospheric oxygen. Although Titan is certainly lifeless—its frigid surface is -290 degrees Fahrenheit—the moon preserves some of the chemical conditions that prevailed on Earth billions of years ago. Titan even has water. The difference is that Earth was warm enough to have liquid water, in which life developed, whereas Titan is so cold that its water froze.

Most of Saturn's other satellites are worlds made chiefly of water ice and lie inside Titan's orbit. These

FACING PAGE: The icy surface of Jupiter's Europa may cover an ocean of water.

FOLLOWING PAGES:
Pages 30-31. Saturn is the solar system's artistic masterpiece.
Page 32. Saturn adorns itself with bright rings.
Page 33. Though only a moon, Saturn's Titan has an atmosphere thicker than Earth's.

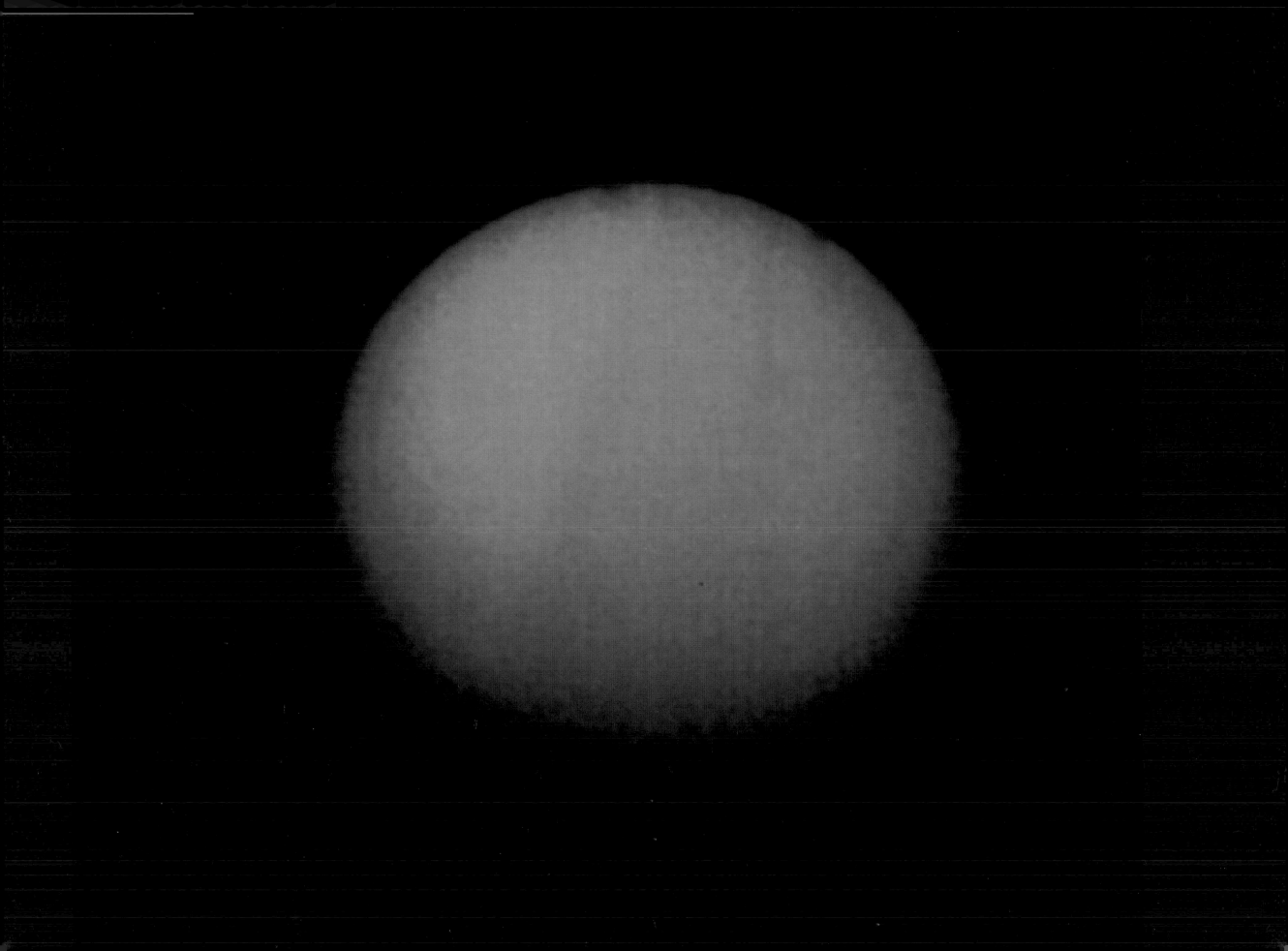

include small "shepherd moons" whose gravitational tugs tend and sculpt some of the rings. Three especially strange moons reside outside Titan's orbit—Hyperion, Iapetus, and Phoebe. Hyperion is a misshapen world that resembles a biscuit rather than a sphere. Iapetus is half white, half black. And Phoebe, which lies much farther out than its peers, is probably a captured asteroid or comet, because it revolves around Saturn backward. Phoebe is dark, and its dust may rain down on Iapetus and darken one side of that moon.

URANUS AND NEPTUNE

The next two planets, Uranus and Neptune, are twin worlds. They have similar natures, and even their discoveries were intertwined. Uranus, twice as far from the Sun as Saturn, is dimly visible to the naked eye. Astronomers had actually seen the green world prior to its discovery, but all mistook it for a star. In 1781, William Herschel, a professional musician and amateur astronomer in England, noticed an object that did not look like a star. Thinking it a comet, he reported the find to professional astronomers, who discovered it to be a planet—the first new planet sighted in the heavens in all of recorded history. It was named for Uranus, the father of Saturn and grandfather of Jupiter.

Almost as soon as Uranus was discovered, however, it spelled trouble, for it sometimes moved too quickly, other times too slowly. Eventually astronomers suspected that a still more distant planet was to blame, a planet whose gravity yanked Uranus and sped it up or slowed it down. English astronomer John Couch Adams and French astronomer Urbain Leverrier independently predicted the unseen world's position, and in 1846 German astronomers Johann Galle and Heinrich d'Arrest found it, close to where Adams and Leverrier had predicted. The planet was blue-green, and it was named Neptune, for the god of the sea.

Once shrouded in mystery—even their rotation

Neptune 17 Earth masses, about one-twentieth the mass of Jupiter. Yet all four giant planets have thick hydrogen-helium envelopes surrounding rock-water cores weighing roughly 10 Earth masses. This similarity points to their greatest difference: on Jupiter and Saturn the hydrogen and helium dominate, whereas on Uranus and Neptune the rock-water core accounts for most of the mass.

Uranus and Neptune have similar colors—Uranus is green and Neptune turquoise blue. On Uranus, the green results from methane gas, which constitutes a small part of its atmosphere. Methane removes red light and leaves green. Neptune's bluer color arises both from methane and from blue particles in its clouds. Both planets spin fast, but Uranus lies on its side as it turns. Perhaps a giant asteroid hit Uranus and knocked it over.

Because Uranus and Neptune reside far from the Sun, they take a long time to revolve. Uranus circles the Sun only once every 84 Earth years, so a human being on Uranus would celebrate at most one birthday. Neptune takes nearly twice as long, completing an orbit every 165 years. Surprisingly, even though Neptune is a billion miles farther from the Sun's warmth, the two planets share exactly the same temperature, −353 degrees Fahrenheit. This similarity stems from one of their few differences: Neptune, like Jupiter and Saturn, radiates more energy than it receives from the Sun, whereas Uranus, for some reason, does not. As a result, Neptune's atmosphere is livelier. When Voyager 2 flew past in 1989, Neptune sported a great dark spot reminiscent of Jupiter's great red spot, but in the 1990s the Hubble Space Telescope showed that this dark spot had vanished.

Both planets have rings, albeit rings so dark they would go unseen by the naked eye of a space trav-

FACING PAGE: The discovery of Uranus doubled the size of the known solar system.

FOLLOWING PAGES:

Page 36. Neptune is the last of the Sun's giant planets.

Page 37. Nitrogen ice sheathes Neptune's moon Triton.

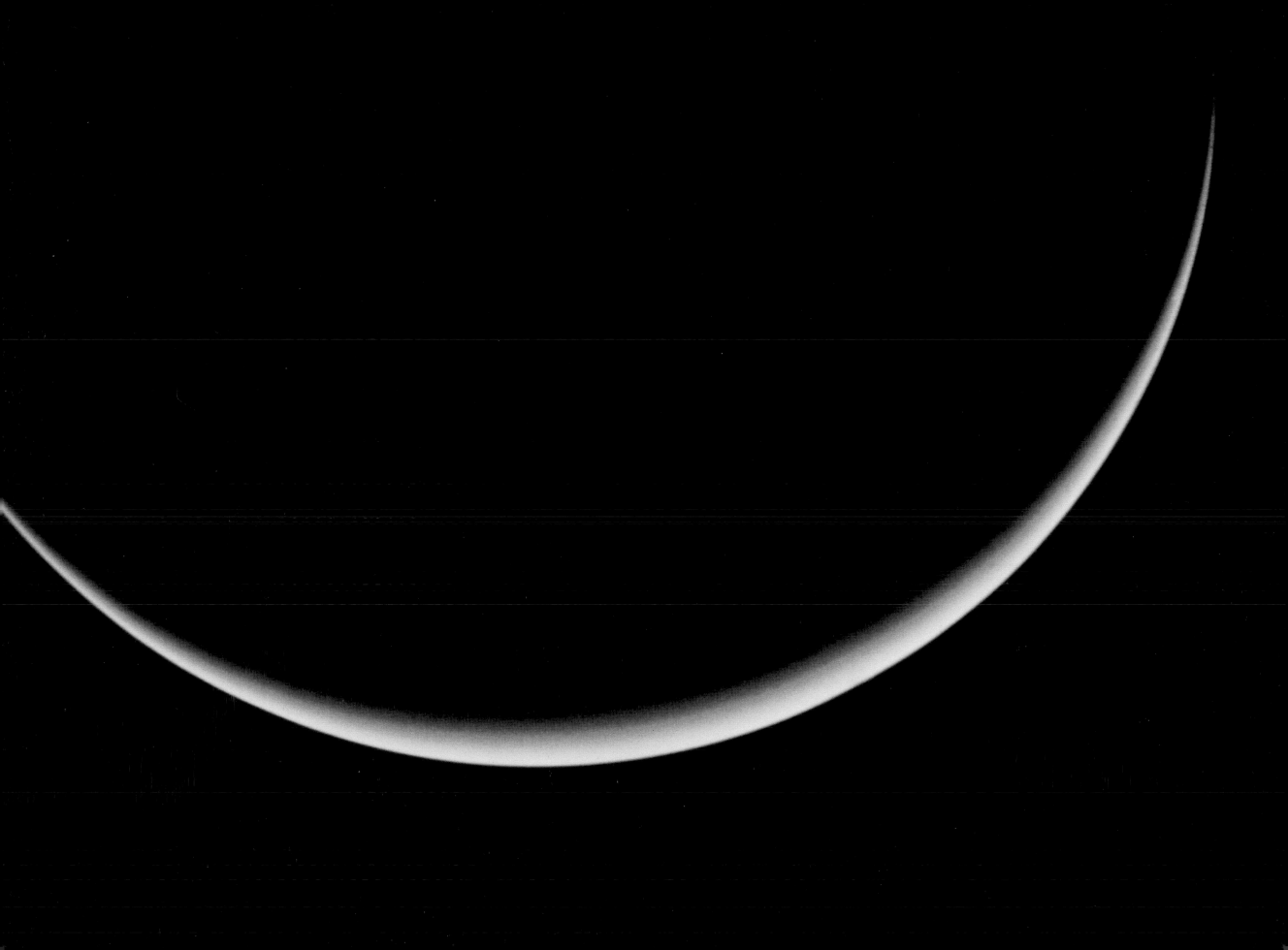

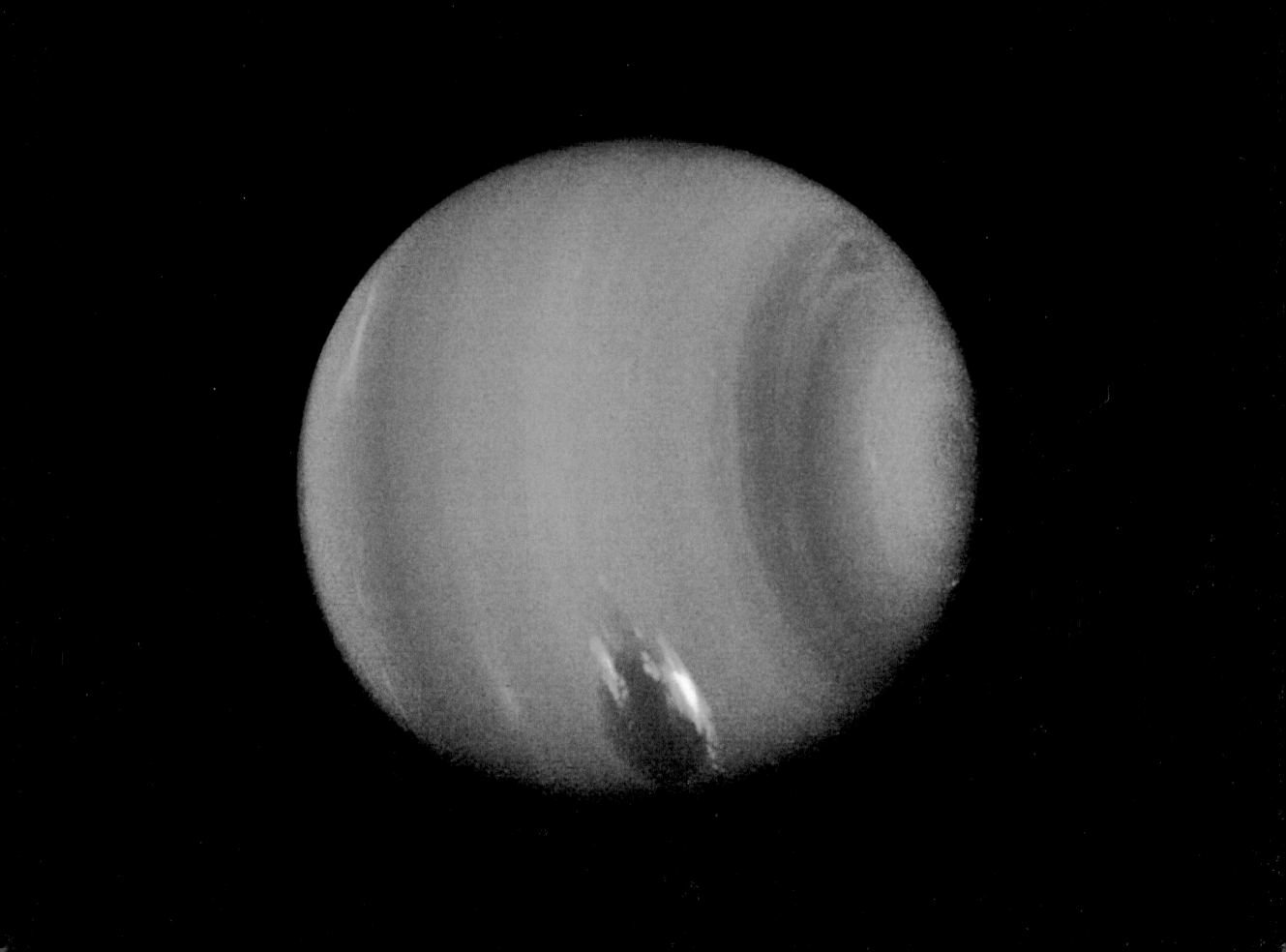

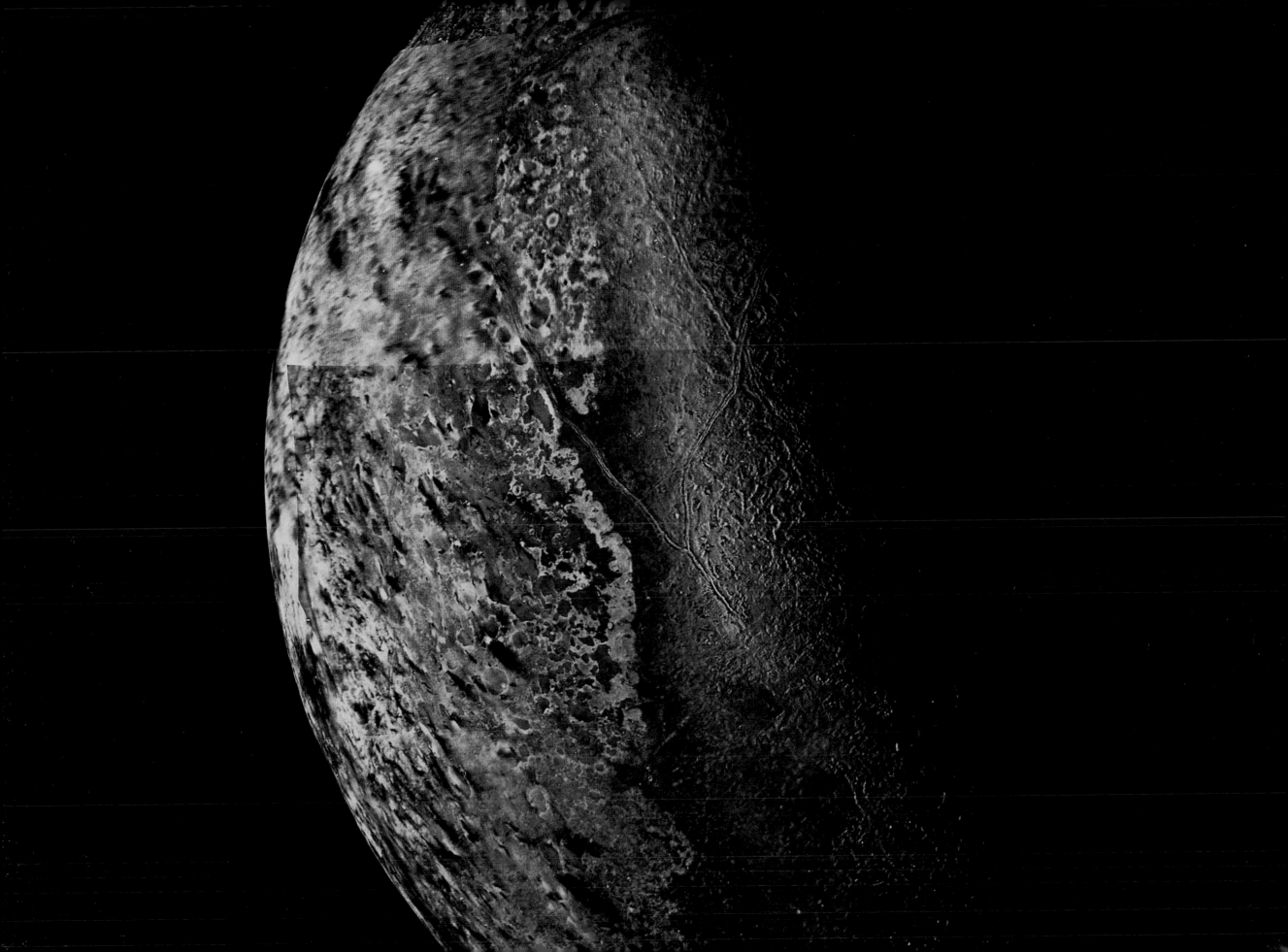

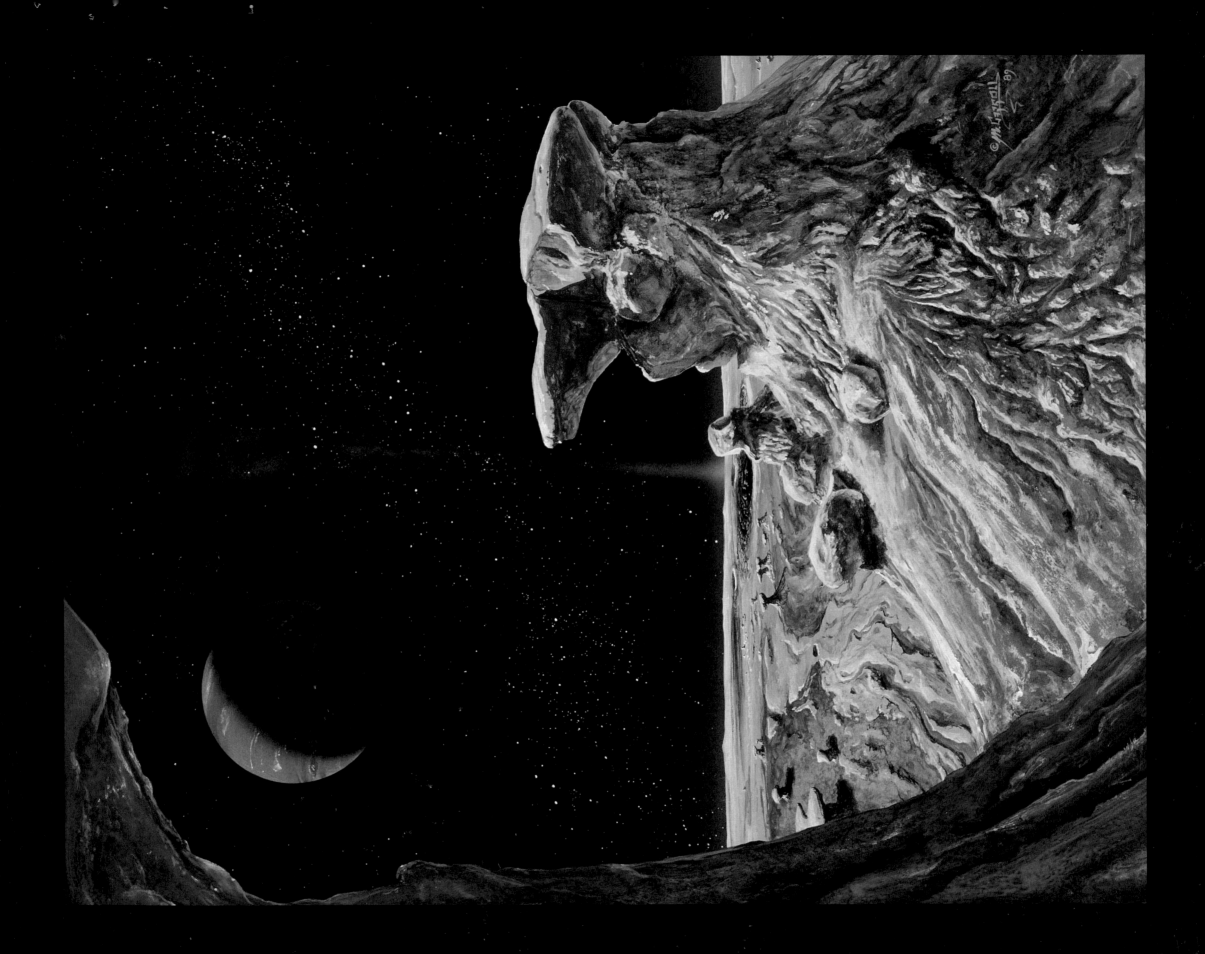

eter flying past them. Uranus's rings were discovered in 1977, when they blocked the light of a star the planet passed in front of. Neptune's rings were found the same way the following decade.

In addition to rings, Uranus has at least eighteen satellites, but none large; the planet is the only giant without a big moon. Neptune's satellite system is only eight moons strong, but the largest falls into the class of big moons: Triton, discovered just two and a half weeks after Neptune itself. Even before Voyager's visit, Triton seemed strange, because it is the only large moon that circles its planet backward, suggesting that it was not born with Neptune but was instead captured by it. Voyager revealed a thin nitrogen atmosphere, an icy surface, complete with a huge polar cap, and geysers shooting material five miles high. Moreover, Triton turned out to be the coldest celestial body any spacecraft has yet measured: –390 degrees Fahrenheit.

The second largest moon is Proteus, which astronomers now realize they had first detected in 1981 when it passed in front of a star; however, not until Voyager saw it did they accept its existence. The third largest Neptunian moon, odd little Nereid, follows the most elliptical path of any known satellite. Discovered in 1949, the moon takes one Earth year to revolve. When nearest Neptune, Nereid lies less than 1 million miles from the planet, but when farthest, the small moon ventures 6 million miles from its master. Nereid's odd orbit may be Triton's fault. Triton probably once roamed around the Sun on its own, but Neptune then snatched it for itself. During this capture, Triton's gravity may

have disturbed the other moons, casting some into Neptune and ejecting others, thus leaving Neptune with fewer moons than other giant planets. In addition, Triton may have stretched Nereid's orbit into a long ellipse.

...HOUGH TRITON is ...w a satellite, a similar ...rld continues to orbit Sun on its own: Pluto

PLUTO

Although Triton is now a satellite, a similar world continues to orbit the Sun on its own: Pluto, outermost of all the Sun's planets, and the last to be found. The astronomer who led the search that would ultimately capture Pluto started his investigations in 1905, a quarter century before the tiny world was actually detected. Percival Lowell, proponent of intelligent life on Mars, thought a ninth planet, Planet X, perturbed Uranus and instigated irregularities that Neptune's existence could not explain. Although Lowell and his assistants searched for many years, he died in 1916 without finding the coveted world. Lowell Observatory resumed the search in 1929, and in 1930 Lowell astronomer Clyde Tombaugh spotted faint Pluto on a photographic plate of the constellation Gemini. But Pluto seemed small, much dimmer than predicted. Lowell had envisioned a world some seven times heavier than Earth, but the real Pluto looked so faint that it might lack the gravitational power to perturb Uranus—even though Pluto had been found during a hunt Lowell had inspired and just a few degrees from one of two places he had pinpointed. Over the following decades, astronomers estimated that Pluto was smaller than Mars but larger than Mercury, making the far-off world the solar system's second smallest planet.

During the 1970s, however, astronomers learned just how tiny Pluto really is. In 1978, James Christy of the U.S. Naval Observatory in Washington, D.C., discovered a moon circling Pluto. This moon feels Pluto's gravity, which derives its strength from Pluto's mass, so the moon's motion around Pluto betrays Pluto's mass

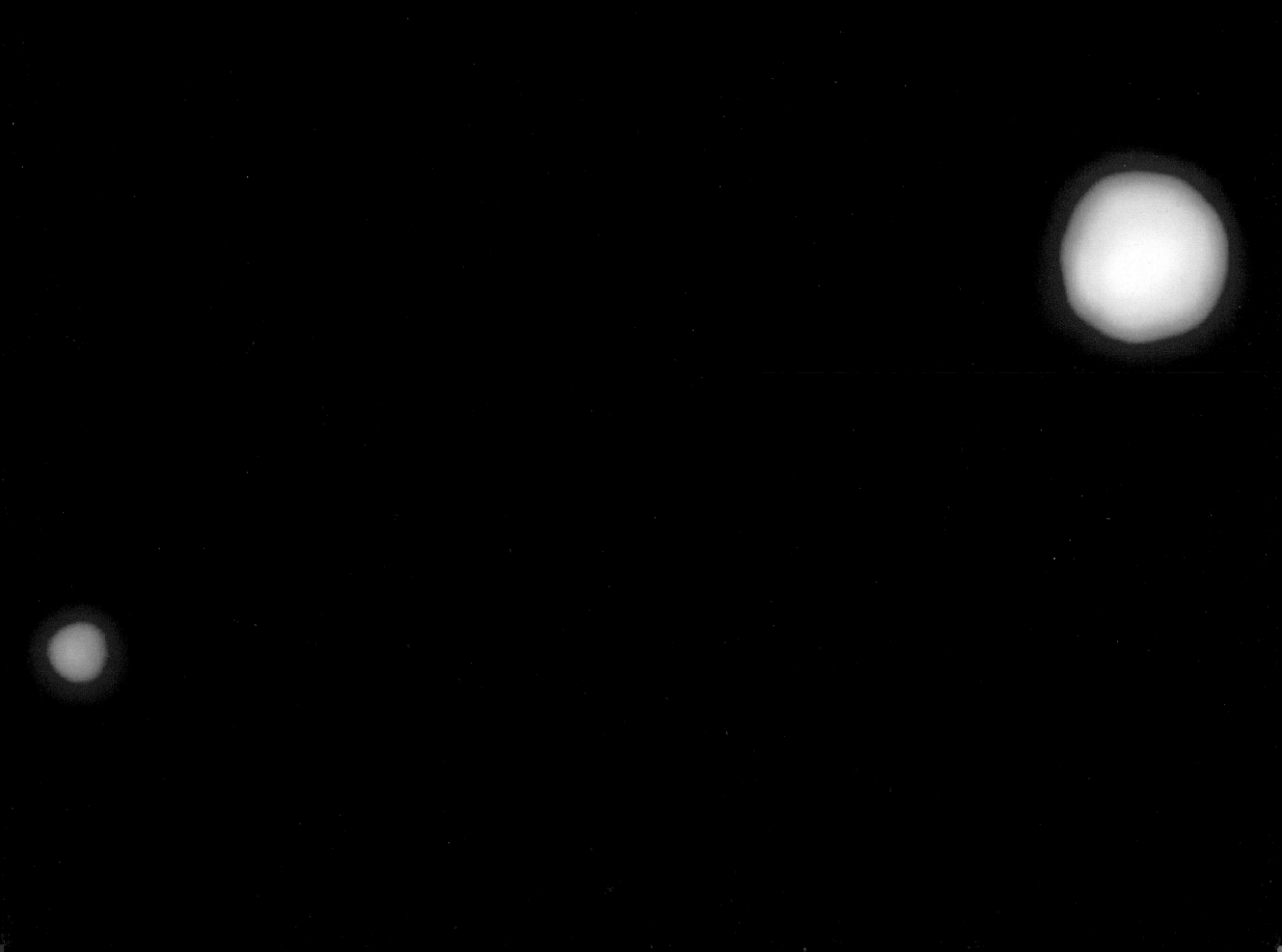

consider it a bona fide planet. It certainly does not affect Uranus significantly. Ironically, the apparent irregularities in Uranus's motion that led Lowell and his observatory to Pluto must have been nothing more than observational error.

Because of Pluto's remoteness, sunlight shines on Pluto with less than a thousandth of its strength on Earth, or a few hundred times brighter than moonlight, so even daytime on Pluto looks fairly dark. Indeed, this darkness partially inspired the planet's name, for Pluto was the god of the underworld, dwelling in darkness. In like fashion, Pluto's moon was christened Charon, after the ferryman who carried souls across the river Styx and into the underworld.

Pluto requires more time than any other planet to orbit the Sun, 248 years. On average, Pluto lies 3.7 billion miles from the Sun, but its orbit is so elliptical that at times Pluto skirts closer to the Sun than Nep-

tune is. Pluto executed its most recent intra-Neptunian excursion between 1979 and 1999; it will not repeat the feat until the twenty-third century. The two planets will never collide, because whenever Pluto equals Neptune's distance from the Sun the two worlds are far apart.

Between 1985 and 1990, astronomers watched Pluto and its moon repeatedly eclipse each other. The duration of these eclipses established the diameters of Pluto and Charon. Pluto is slightly smaller than Triton, and Charon is about half Pluto's size, giving Pluto and

Charon the largest moon-to-planet size ratio in the solar system. For comparison, the Moon is a quarter of the Earth's diameter. Charon is so large and so close to Pluto that from the planet's surface the moon would look seven times wider than the Moon does from Earth.

Just as the same side of the Moon always faces the Earth, so the same side of Charon always faces Pluto. In addition, the same side of Pluto always faces Charon, so the moon never rises or sets; it simply hovers over the same part of Pluto, and a Plutonian would see the moon forever lodged at the same altitude above the horizon. Because one side of Pluto faces away from Charon, that hemisphere never sees the moon at all.

Like Neptune's Triton, Pluto has a nitrogen atmo-

sphere. It was first detected in 1988, when Pluto passed in front of a star whose light faded gradually rather than abruptly, signaling a gaseous envelope around Pluto. The atmosphere's composition was inferred in 1992, when astronomers discovered that Pluto's surface had nitrogen ice, some of which must vaporize and become gas. Thus, far-off Pluto is the only other planet whose main gas is the same as Earth's.

Pluto and Triton have similar densities, indicating similar compositions—about 75 percent rock and 25 percent ice. This rock-rich composition suggests that both Pluto and Triton formed far from any planet, where it was cold. At low temperatures, oxygen binds with carbon to make carbon monoxide (CO), leaving little oxygen to join with hydrogen and form water (H_2O); so Pluto and Triton were born with relatively little water ice. In contrast, had they formed around a giant planet, they would have bathed in the newborn giant's warm glow. Warmth converts carbon monoxide into methane (CH_4) and liberates the oxygen, which combines with hydrogen to create water. Thus, most satellites of the giant planets, which formed around their planets, possess greater proportions of water ice than Pluto and Triton do. Although Pluto and Triton had similar origins, fate intervened: Triton was captured by Neptune, while Pluto remained free.

> **FAR-OFF PLUTO is the only other planet whose main atmospheric gas is the same as Earth's**

FACING PAGE: Pluto and its moon Charon patrol the solar system's frontier.

14,000 years and Hale-Bopp in 2,400 years. The one exception to the rule is Halley, which swings around the Sun every 76 years, similar to a human lifetime; indeed, Mark Twain was born during one passage and died during the next. Short-period comets other than Halley fail to impress, having swung by the Sun so many times that they have exhausted much of their tail-generating ice. Several asteroids pursue highly elliptical, cometlike orbits and may be burned-out comets.

ONLY ABOUT ONCE a decade does a great comet appear—so bright and beautiful that it captivates the public

The solar system does not end with Neptune and Pluto. Beyond their orbits lie trillions of comets, ice-bearing bodies that hibernate in the extreme cold of the outermost solar system. If one falls toward the Sun, however, sunlight warms the comet and vaporizes some of its ice, sparking a beautiful tail. Contrary to the impression it creates, the tail does not necessarily trail the comet. Instead, the tail points away from the Sun, because the solar wind blows cometary gas and dust away.

Most comets that astronomers can see travel along highly elliptical orbits around the Sun. For example, Comet Halley (rhymes with "valley") darts from beyond Neptune to within Venus's orbit. Comets pass Earth all the time, but only about once a decade does a *great* comet appear—so bright and beautiful that it captivates the public. A comet's greatness depends on several factors, including its size (the bigger, the better); how close it gets to the Sun (the closer, the better, because the more the ice vaporizes); and how close the comet comes to Earth (the closer, the better). Sometimes the same comet on different passes can be both great and not so great. Halley's Comet was great in 1910, when it came so close that the Earth passed through its tail, but was nothing special in 1986, when it passed far from Earth. Its next return, in 2061, should be better.

The 1990s blessed Earth with two great comets. After twenty years of great-comet drought, Comet Hyakutake skirted by Earth in 1996. It was small but came so close to Earth that in dark skies it looked spectacular. During 1997, a comet passed far from Earth but was so large—20 miles across, versus Halley's 5 x 5 x 10 miles—that it amazed even people who lived among bright city lights. This was the great Comet Hale-Bopp, which remained visible to the unaided eye longer than any other comet of the twentieth century.

Despite their dramatically different appearances, asteroids and comets are really two sides of the same coin: both are objects left over from the formation of the solar system, objects that did not get swallowed by the Sun or planets. Unlike asteroids, comets bear ice, because comets arose in the cold of the outer solar system. Yet both asteroids and comets are fossils of the solar system's birth. Similar objects smashed together to form the Earth and other planets.

FACING PAGE: Comets, such as Hyakutake, carry ice that vaporizes in sunlight.

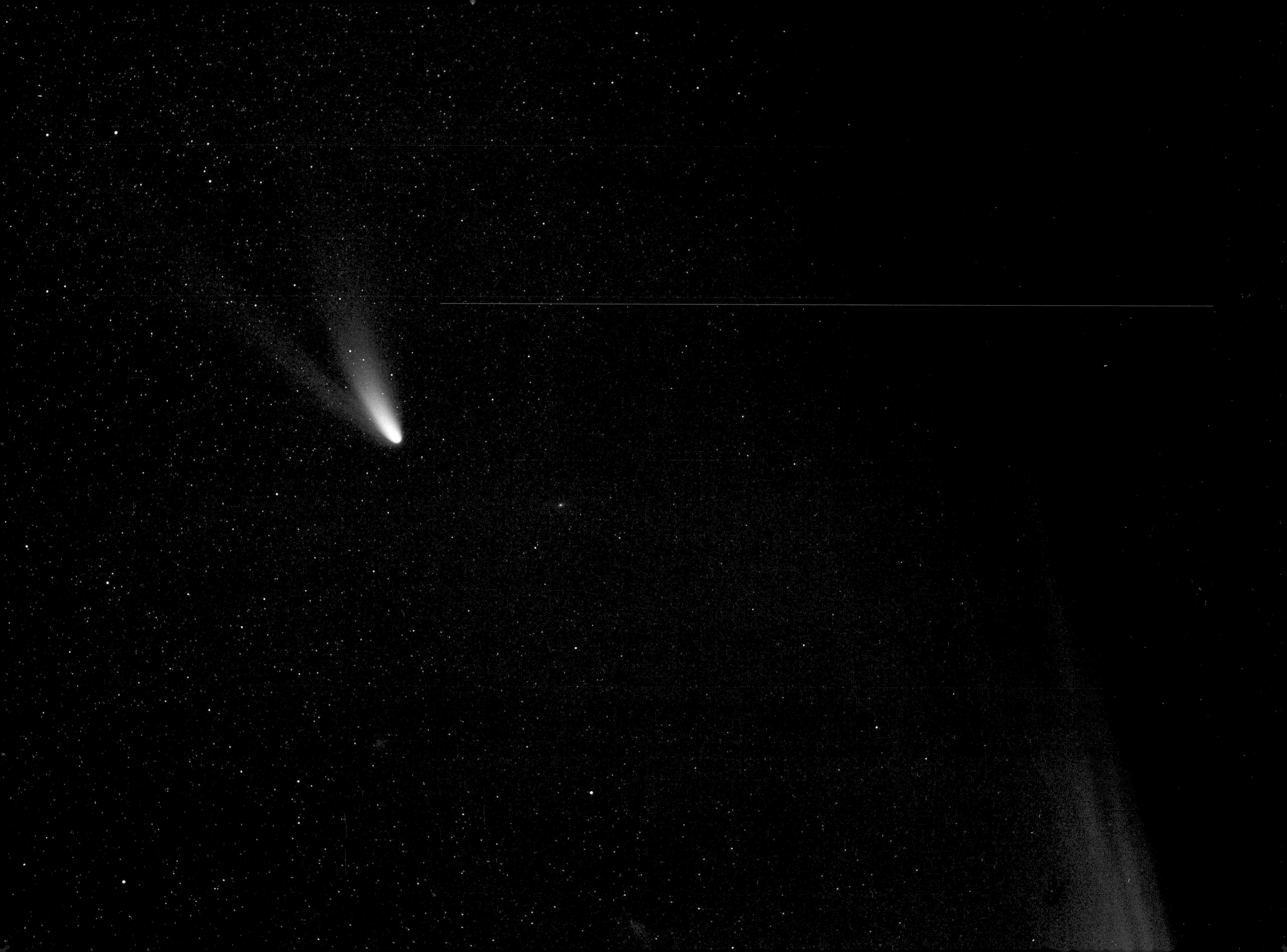

Short- and long-period comets originate from different regions of the outer solar system. The first region, just past the orbit of Neptune, supplies most short-period comets and is often called the Kuiper belt, after prominent American astronomer Gerard Kuiper, who proposed its existence in 1951. However, an obscure Irish astronomer, Kenneth Edgeworth, had earlier suggested the same idea, so this zone might better be called the Edgeworth-Kuiper belt. Because the Edgeworth-Kuiper belt is flat, short-period comets usually travel near the same plane as the planets. In 1992, astronomers began detecting objects in the Edgeworth-Kuiper belt. Pluto and Triton may simply be the largest Edgeworth-Kuiper belt members.

ASTEROIDS AND COMETS are really two sides of the same coin: both are objects left over from the formation of the solar system

Edgeworth-Kuiper belt members can migrate inward, toward the Sun. In 1977, American astronomer Charles Kowal discovered the first example, between the orbits of Saturn and Uranus. Kowal called it Chiron, after the son of Saturn and grandson of Uranus, a name that fit the object's odd location. The name turned out to be even more appropriate, however, because Chiron was a centaur, half man and half horse, and Chiron the celestial object is part asteroid and part comet. Chiron was first catalogued an asteroid, but none had ever been seen so far from the Sun, and in 1988 astronomers saw Chiron brighten as its ice began to vaporize—the same behavior comets exhibit. Yet Chiron is much larger than other comets. It is about 120 miles across, six times larger than Hale-Bopp and nearly twenty times bigger than Halley. In 1992 astronomers discovered another object like Chiron; it spends most of its time between the orbits of Saturn and Neptune. Since then, other "centaurs" have emerged, icy bodies that cut across the orbits of the outer planets, likely refugees from the Edgeworth-Kuiper belt.

Far beyond the Edgeworth-Kuiper belt lies another cometary reservoir, the remote Oort cloud, which stretches about two light-years from the Sun, halfway to the nearest star. Because of its much greater distance, astronomers have never observed objects lodged in the Oort cloud. It houses the long-period comets, such as Hyakutake and Hale-Bopp. Unlike short-period comets, these comets come from all directions, above and below the plane of the solar system, because unlike the flat Edgeworth-Kuiper belt, the Oort cloud is roughly spherical.

Whatever its origin, a speeding comet sprinkles space with cast-off dust and ice. Asteroids also splatter material, when other asteroids hit them, and this asteroidal material can be larger and tougher than cometary debris. If the Earth runs into a bit of interplanetary flotsam, called a meteoroid, friction with the atmosphere burns it up in a streak of light called a meteor or shooting star. Countless meteoroids strike the Earth's atmosphere each day, but most are small and never survive their fiery passage to the ground. Larger meteoroids produce bright meteors called fireballs. Those large and tough enough to reach the surface originate primarily from asteroids; after landing, they are called meteorites.

Because comets crisscross the solar system, so do the meteoroid streams they spawn, which follow the comets' orbits. When the Earth encounters one of these streams, large numbers of meteoroids smash into the atmosphere and cause a meteor shower. Debris from Halley's Comet, for example, triggers two meteor showers, the Eta Aquarids in May and the Orionids in October. Meteoroids in a stream travel on nearly parallel paths through space, so all meteors in a particular shower appear to radiate from a single point, just as parallel railroad tracks appear to converge in the distance. A meteor shower is normally named for its radiant: the Orionids, for instance, appear to emanate from the constellation Orion. Still, Orionid meteors streak across other constellations; it's just that the path of any Orionid meteor, if traced back, will pass over Orion.

The best meteor showers are the Perseids in August and the Geminids in December, which produce about sixty meteors an hour, or one a minute. The Perseids have been seen for nearly two thousand years. They arise from Comet Swift-Tuttle, which revolves around the Sun every 130 years and last passed us in 1992, replenishing the supply of meteoroids near the Earth. The Geminids are more recent, having first appeared

dictable meteor shower strikes in November. A product of the unspectacular Comet Tempel-Tuttle, the Leonids normally display only a dozen meteors an hour, but in some years—such as 1799, 1833, 1866, and 1966—they have erupted into titanic meteor *storms* that shot thousands of meteors across the sky in a single hour. Potential outbursts occur every 33 years, the orbital period of Comet Tempel-Tuttle.

A METEOR FLASH is really a miniature reenactment of Genesis

Spectacular though meteor storms are, the meteoroids that generate them are the most minuscule of the many objects that circle the Sun. Yet they, along with asteroids and comets, are remnants of the solar system's birth: a meteor flash is really a miniature reenactment of Genesis, the process that formed the Earth and the other planets circling the Sun.

But the Sun is only one star in a galaxy full of stars, many of which may also have solar systems full of planets, asteroids, comets, meteoroids—and life. To these other stars we now turn.

FACING PAGE: A Geminid meteor flashes across the constellation

heavens sparkle

est outshine the faintest a trillion times over, and different stars glow blue, white, yellow, orange, and red. Look over here and see a massive blue star burst from its birthplace burning brilliantly, while over there, a dim red star struggles simply to ignite its meager nuclear flame. Over here, watch two bright stars, one orange, the other blue, dance around each other gracefully, while over there a dim white dwarf, the burned-out core of a once-Sunlike star, barely pierces the darkness. Over here, witness the ruins of a spendthrift star that exploded and then collapsed into a black hole; over there, nine planets faithfully attend a long-lived yellow star in the Orion arm of the Galaxy.

From Earth, the unaided eye can see several thousand stars, all part of our Galaxy. Yet the Milky Way is so huge that for every star visible to the unaided eye, the Galaxy possesses millions more, giving it a grand total of hundreds of billions of stars—some bright, most faint, but all partners in a colossal system held together by gravity.

FACING PAGE: Like its namesake, the California Nebula seems to be its own nation.

FOLLOWING PAGES:

Pages 50–51. Stars decorate the night and forge life-giving elements like carbon and oxygen.

Pages 52–53. The Horsehead Nebula is a cloud of interstellar gas and dust that resembles a cosmic chess piece.

cent dust particles. The two go together, so that space-bearing thick gas also bears thick dust. The most common interstellar element is hydrogen, the most common in the universe. Helium runs second, offering one atom for every ten of hydrogen. Helium doesn't do much, however, since it doesn't join with other atoms to form molecules. Elements heavier than helium are much rarer, accounting for only one or two atoms—mostly oxygen and carbon—for every thousand of hydrogen. Thus, stars have altered the universe's composition only slightly; the dominant elements remain hydrogen and helium.

Without stars, there would be no life. Stars supply life with light and heat. The Sun warms the Earth and keeps most of its water liquid. Sunlight also triggers the photosynthesis that most plants use to produce energy, energy that plant-eating animals then receive. Stars have helped life in another way, by forging most of the elements in the cosmos. Before stars shone, the universe had only hydrogen and helium, the two lightest elements, and a bit of lithium, the third lightest element; but these three elements can't construct the complex molecules that life needs. Instead, life requires heavier elements, like carbon, nitrogen, oxygen, and iron, which stars create from lighter elements and then spew into space when they die. The oxygen you breathe, the calcium in your bones, and the iron in your blood were all manufactured by stars that died billions of years ago. Most every atom on Earth has a heritage in one star or another.

THE INTERSTELLAR MEDIUM

The story of the stars actually begins in the space between them, the interstellar medium, which bears the material that spawns new stars and to which they contribute when they die. In this way, stars are to the interstellar medium as trees are to soil: a tree is born from the soil and nourishes it by shedding leaves and branches into the soil, which then produces more trees.

The space between the stars looks empty, which is what it mostly is. A cubic centimeter of terrestrial air packs 25 million million million molecules, but the same volume of interstellar space in the Galactic disk typically holds just a single atom—by laboratory standards, a perfect vacuum. However, the Galaxy is so big that if you added up all this tenuous material, its total mass would equal 5 to 10 billion stars like the Sun—approximately 5 to 10 percent of the mass of the Galactic disk.

MOST EVERY ATOM on Earth has a heritage in one star or another

Interstellar hydrogen takes two main forms, atomic and molecular. Atomic hydrogen exists as individual hydrogen atoms. In its neutral form, it is called H I (pronounced "H one") and emits 21-centimeter-long radio waves that astronomers first detected in 1951. H I conglomerates in cool clouds that shiver at 50 Kelvin (−370 degrees Fahrenheit), measure a couple of dozen light-years across, and contain 10 to 20 atoms per cubic centimeter. Atomic hydrogen makes up roughly two-thirds of interstellar hydrogen, but it rarely if ever gives birth to stars. Stars form when interstellar material collapses under the weight of its gravity, and H I clouds are too fluffy to collapse.

FACING PAGE: The Orion Nebula has spawned thousands of new stars.

In contrast, the other main type of interstellar hydrogen is a more fertile medium for forming stars: molecular hydrogen (H_2), two hydrogen atoms joined together. Unlike atomic hydrogen, molecular hydrogen does not advertise its presence through radio broadcasts, so astronomers track it by observing a radio-loud molecule that accompanies it, carbon monoxide (CO). This deadly gas, present in car exhaust and cigarette smoke, contains the two most common reactive atoms after hydrogen and outnumbers all other interstellar molecules but molecular hydrogen. Even so, only one carbon monoxide molecule exists for every ten thousand hydrogen molecules. Astronomers first picked up carbon monoxide's radio waves in 1970. Clouds bearing H_2 and CO also contain smaller amounts of more complex molecules, such as water, ammonia, alcohol, and formaldehyde.

By interstellar standards, molecular clouds are dense. They must be, because molecules are fragile and space is harsh, filled with deadly ultraviolet radiation that rips molecules apart, so interstellar molecules

A SINGLE molecular cloud might span a hundred light-years

can survive only where gas and dust are dense enough to shield the molecules. Molecular clouds are therefore the densest interstellar material—so dense that their dust can block starlight and produce dark nebulae such as the Horsehead Nebula, which looks like an eerie cosmic chess piece in Orion, and the Coalsack Nebula in the Southern Cross. (In Latin, *nebula* means "cloud"; the plural is *nebulae*.) Molecular clouds harbor hundreds, thousands, or millions of molecules per cubic centimeter, and they are cold, sometimes just 10 Kelvin (−440 degrees Fahrenheit). The high density means the gas can collapse under its own weight, and the cold temperature means thermal pressure won't stop the collapse, so gravity wins and stars arise.

Atomic and molecular clouds rule different parts of the Galaxy. The Sun lies roughly halfway from the center of the Milky Way to the edge of the disk, so the Sun marks a convenient dividing point: astronomers often call that part of the Galaxy which lies closer to the center than the Sun the inner Galaxy and that part which lies farther the outer Galaxy. Atomic hydrogen, the form that does not give birth to new stars, prevails in the outer Galaxy, while the star-breeding molecular hydrogen dominates the inner Galaxy. About 70 percent of the Galaxy's atomic hydrogen lies farther from the Galactic center than the Sun, and some even extends beyond the Milky Way's disk of stars. In contrast, about 90 percent of the Galaxy's molecular hydrogen lies closer to the Galactic center than the Sun. Most stars therefore form in the inner Galaxy.

Molecular clouds come in different sizes. Most stars are born in the largest clouds, called giant molecular clouds, so chances are the Sun was too. A single giant molecular cloud might span over a hundred light-years, harbor hundreds of thousands of solar masses, and spawn over a thousand stars. The nearest giant molecular cloud is 1,500 light-years from Earth, in the constellation Orion. A small part of this giant molecular cloud glows from the radiation of hot stars. This is the famous Orion Nebula, which even the unaided eye can see as a fuzzy patch of light in Orion's sword. Giant molecular clouds give birth to stars large and small. In contrast, smaller molecular clouds, some of which lie 500 light-years away in the constellation Taurus, do not produce the largest stars.

FACING PAGE: Just a few hot stars set the entire Orion Nebula aglow.

boring hotter stars sometimes combine ionized hydrogen's red glow and dust's reflected blue to paint space magenta or purple.

Aside from coloring and darkening the sky, interstellar dust plays several other roles. First, by blocking starlight, it cools molecular clouds, which helps stars form, since cooler clouds more easily collapse. Second, dust protects molecules from the radiation that would destroy them. Third, dust allows molecular hydrogen to arise in the first place, because hydrogen atoms do not join together in space. Dust particles resemble interstellar singles bars that pair up lonely hydrogen atoms: a hydrogen atom settles on a dust grain, finds another hydrogen atom there, and a marriage is made. And fourth, dust grains circling a newborn star stick to other dust grains and can ultimately grow into planets.

DUST GRAINS can ultimately grow into planets

Interstellar hydrogen can also assume a form other than neutral atomic or molecular, one that emits a beautiful red glow which lights many star-forming regions. The red light actually signals the presence of blue stars. If an interstellar cloud makes a star exceeding twelve times the Sun's mass, the star becomes so hot and blue it radiates extreme ultraviolet light, which tears electrons from hydrogen atoms and thereby ionizes the hydrogen. When the electron rejoins the atom, it can send out red light. Just as neutral atomic hydrogen is called H I, ionized hydrogen is called H II ("H two"). The most famous H II region is the Orion Nebula, set aglow by several hot young stars in the Orion molecular cloud. Other H II regions include the North American Nebula in the constellation Cygnus, the Rosette Nebula in Monoceros, and the Eagle Nebula in Serpens. However, hot, high-mass stars that ionize hydrogen are rare, so some star-forming regions, such as the one in Taurus, do not glow red.

Tiny dust particles also sprinkle space, particles that form primarily in the atmospheres of cool, aging stars called red giants. Dust makes up only 1 percent of the interstellar medium's mass, which may not sound like much, but if Earth's air held the same percentage of dust, you wouldn't be able to see your feet. Interstellar dust scatters and absorbs light, so dense clouds like the Horsehead Nebula and the Coalsack Nebula look black. Dust's dimming of starlight is called interstellar extinction.

Dust also reddens starlight, which is called interstellar reddening. This happens because red light penetrates dust better than blue light does, so a yellow star emitting equal amounts of red and blue light looks red if it lies behind dust. A similar phenomenon reddens the Sun at sunrise and sunset, when we view the Sun through the large amount of air on the horizon. Dust also reflects starlight. For example, the blue stars in the Pleiades cluster are too cool to ionize hydrogen and make it glow red, but dust reflects their light and shrouds the Pleiades in blue nebulosity. Regions har-

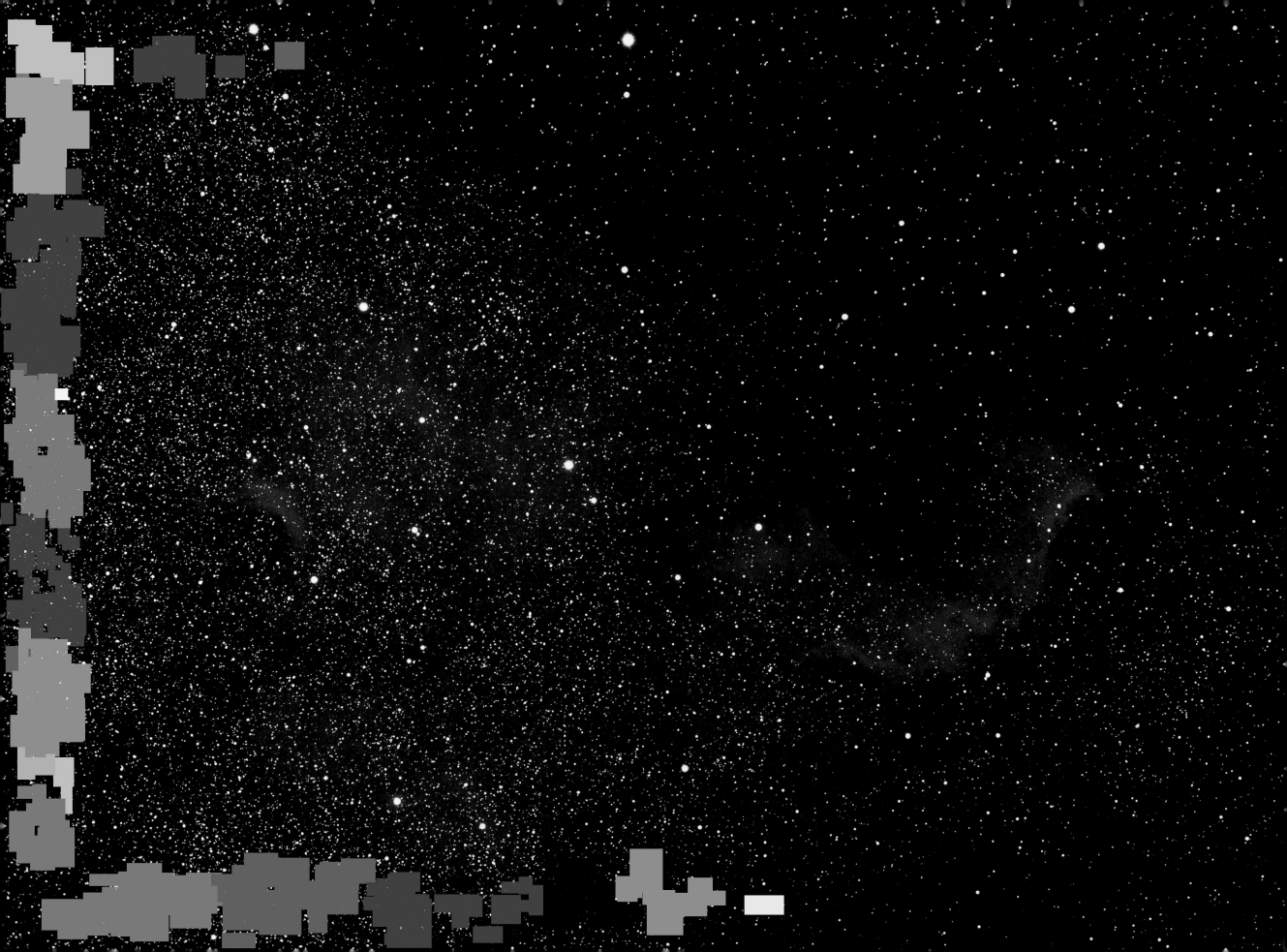

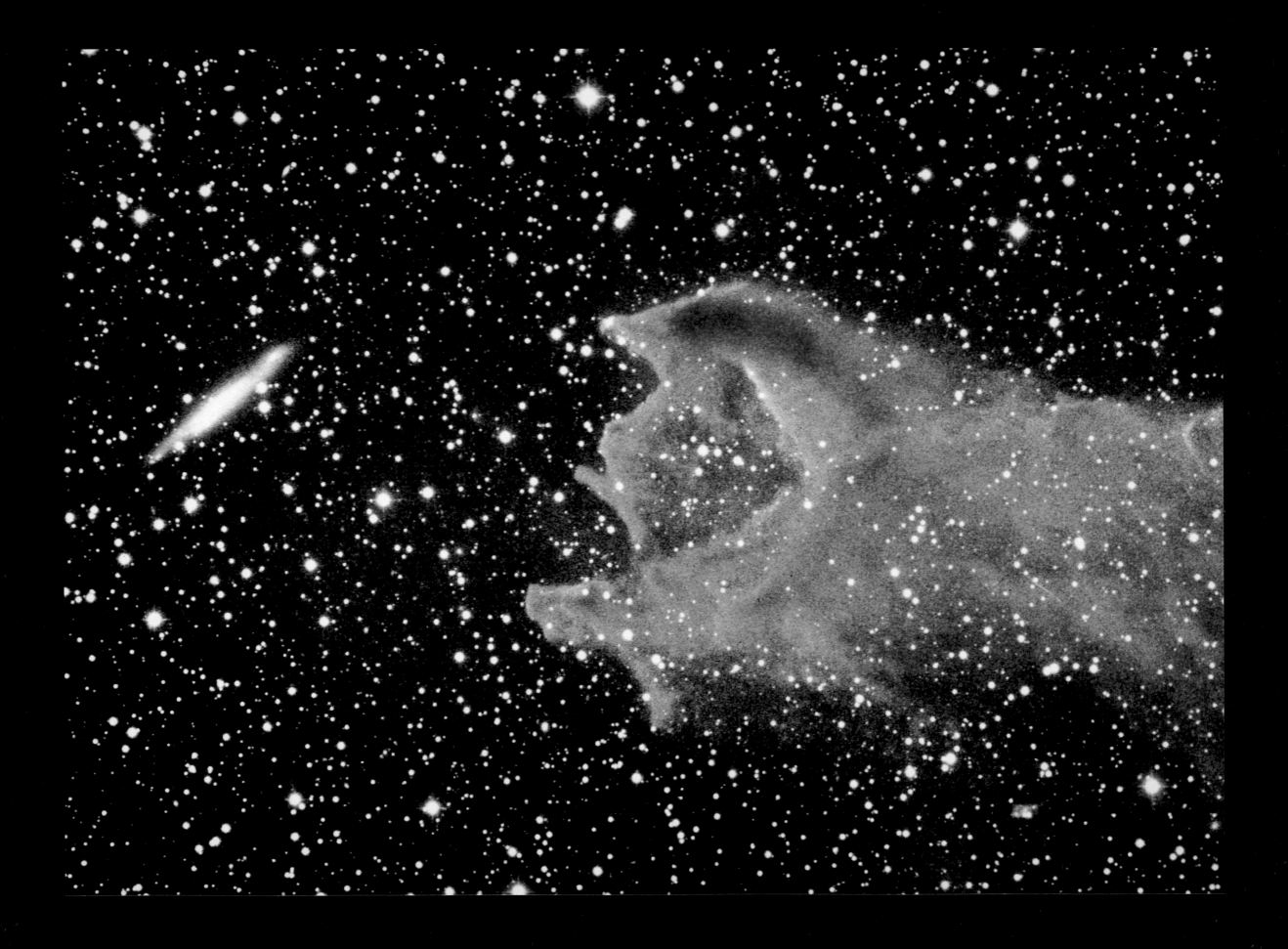

The Milky Way Galaxy is a vigorous star creator. Each year our Galaxy gives birth to about ten new stars—more than all the other galaxies of the Local Group produce together. These star births take place in the Galaxy's spiral arms, where interstellar gas gets squeezed so that it becomes dense and molecular.

Inside molecular clouds, dense cores a few light-years across give birth to individual stars. The core collapses, perhaps under its own weight, or perhaps because an external shock, such as the blast of an exploding star or pressure from stellar radiation, triggers it. Whatever the exact cause, the core collapses from the inside out: first material near the center falls inward, followed by material farther out. This material crashes into the star-to-be and releases energy, which warms the object and makes it glow—a new star is born. Unlike the Sun, it is powered by gravitational energy, not nuclear energy, because gravity sets the material into motion, just as gravity forces water to fall over a hydroelectric dam and generate electricity. At first, observers on Earth cannot see the star at visible wavelengths, because the gas and dust in the outer part of the cloud core block the view. Eventually the star breaks through its cocoon, and its light reaches Earth.

For stars with masses like the Sun's, this phase corresponds to the T Tauri stage, named for a flickering young star in Taurus that converts gravitational energy into light. T Tauri stars range in age from half a million to 3 million years old. They are larger than the Sun and so shine more brightly. As the T Tauri star contracts, its core heats up and soon starts to transform hydrogen into helium—the same nuclear reaction that powers the Sun and most other stars. The star is then said to be a main-sequence star.

Stars are usually born with many others. The most famous star formation site is the Orion Nebula, which nurses thousands of newborn stars. If a molecular cloud creates high-mass stars, as the Orion Nebula

views of three "star pillars," molecular clouds that ar being assaulted by ultraviolet radiation from hot youn stars near them. Some newborn stars, still attached t the pillars by fingers of molecular gas, look like the flames at the ends of narrow candles.

In many cases, astronomers believe, more tha just a star is born; planets also arise, because when th gas and dust collapse, not all of it falls into the centra object. Instead, some material dives in with enoug orbital speed that it circles the star and forms a swirlin disk of gas and dust. Inside this disk, dust particles stic together and grow, eventually becoming asteroids an comets that collide and merge to become planets.

EACH YEAR our Galaxy
gives birth to about ten
new stars

FACING PAGE: The Eagle Nebula gives birth to new stars in Serpens.

FOLLOWING PAGES:

Page 74. Ultraviolet rays from young stars assault columns of gas and dust in the Eagle Nebula, unveiling new stars.

Page 75. A cluster of young stars emerges from interstellar material.

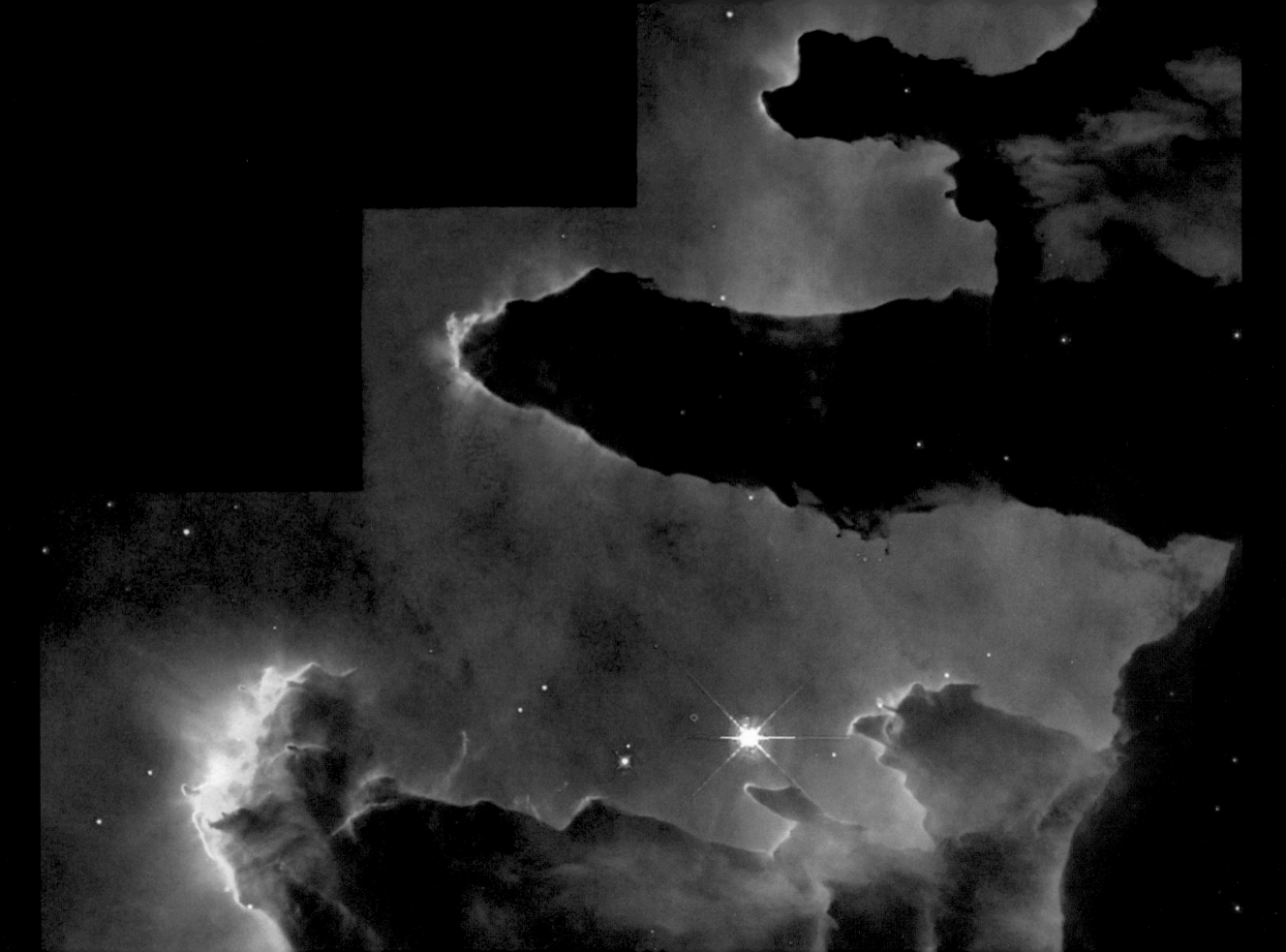

STAR LIFE

All stars are not created equal. A star's life and destiny depend primarily on how much mass it is born with.

Some stars start life with a hundred times more mass than the Sun, but most are born with just a fraction of a solar mass. This wide range of mass explains some of the great diversity stars exhibit. Stellar properties also depend on evolution. For example, a star born with the Sun's mass can be bright and yellow, very bright and red, or dim and white–all depending on the star's age. Mix stars of different masses, all in different stages of evolution, and one ends up with an enormous zoo: big red stars, little white stars, bright yellow stars, very bright blue stars, dim red stars, modest orange stars, and invisible black holes.

To make sense of this diversity, astronomers use the Hertzsprung-Russell diagram, named for Danish astronomer Ejnar Hertzsprung and American astronomer Henry Norris Russell, who independently formulated it early in the twentieth century. The Hertzsprung-Russell diagram sorts the stars by placing different varieties into different areas of the diagram.

For all its power, the H-R diagram is simple. It plots a star's intrinsic brightness, or luminosity, against its temperature. On the H-R diagram, bright stars reside at the top, faint stars at the bottom; hot stars on the left, cool stars on the right. Thus, a star that is bright and hot like Rigel would be placed in the upper left of the H-R diagram, whereas a star that is faint and cool like Proxima Centauri would appear in the lower right.

Luminosity, the vertical axis of the H-R diagram, is a basic stellar property: the amount of light the star emits into space. Different stars radiate radically different amounts of light. The Sun lies midway between the luminosity extremes–the brightest stars shine a million times more brightly, the dimmest a million times more faintly–but the Sun is *not* an average star, because stars dimmer than the Sun far outnumber stars brighter.

In order to compute a star's luminosity, astronomers must know not only how bright the star looks but also how far the star is from Earth. After all, a distant star will look faint no matter what its luminosity. Luminosity is often called intrinsic brightness, since it depends only on how powerfully the star shines, as opposed to apparent brightness, which depends partly on factors unrelated to the star's power, such as its distance from Earth.

A star's surface temperature, the horizontal axis of the H-R diagram, is also a vital stellar property. Temperature relates to color, exactly opposite from the way one might think. Hot stars are blue and white; warm stars, like the Sun, are yellow; and cool stars are orange and red. Ordinary experience teaches the reverse: colors of the sea are considered "cool" and colors of fire "hot." Physically, however, blue corresponds to hot

ALL STARS are not created equal

temperatures, red to cool. Heat a metal rod and it first glows dull red. Then, as it becomes hotter, it turns orange, then yellow, then white. On the H-R diagram, the hot blue and white stars appear on the left, the cool orange and red stars on the right.

A star's temperature affects its spectrum, the light the star emits at various wavelengths. Different atoms and molecules absorb light of different wavelengths and imprint themselves as dark lines on the spectra of stars having different temperatures. For example, helium is normally very stable, its two elec-

-trons strongly bound to the nucleus. However, blue stars are so hot they jostle these electrons and thus exhibit strong spectral lines of this element. Red stars also have helium in their atmospheres but are too cool to disturb its electrons, so helium lines do not appear in these stars' spectra.

Using such spectral differences, astronomers classify stars into seven main spectral types, which correlate with stellar temperature and color. Each spectral type is marked by a different letter. From hot and blue to cool and red, they are O (blue), B (also blue), A (white), F (yellow-white), G (yellow, like the Sun), K (orange), and M (red). The mnemonic: Oh, Be A Fine Guy/Girl, Kiss Me!

When Hertzsprung and Russell plotted stars of different luminosities and temperatures on the H-R diagram, they found a startling thing: stars do not scatter randomly over it. Instead, they congregate in some areas of the H-R diagram and avoid others. The H-R

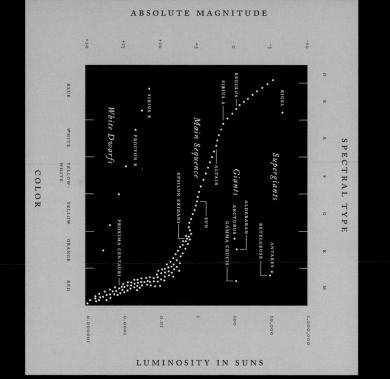

ABSOLUTE MAGNITUDE

SPECTRAL TYPE

O B A F G K M

COLOR

BLUE WHITE YELLOW-WHITE YELLOW ORANGE RED

LUMINOSITY IN SUNS

1,000,000 10,000 100 1 0.01 0.0001 0.000001

RIGEL
REGULUS
SIRIUS A
SIRIUS B
PROCYON B
White Dwarfs
Main Sequence
Supergiants
Giants
ALTAIR
SUN
EPSILON ERIDANI
PROXIMA CENTAURI
ANTARES
BETELGEUSE
ALDEBARAN
ARCTURUS
GAMMA CRUCIS

-10 -5 0 +5 +10 +15 +10

diagram reveals and segregates three chief varieties of stars: main-sequence stars; giants and supergiants; and white dwarfs.

The modern H-R diagram shows that 95 percent of stars reside on the main sequence, a diagonal band swooping from upper left (bright and blue) to lower right (faint and red). It spans all colors—blue, white, yellow, orange, and red—and most luminosities. The Sun is a main-sequence star, as are Regulus (a blue B star), Sirius (a white A star), Tau Ceti (a yellow G star, like the Sun), Epsilon Eridani (an orange K star), and Proxima Centauri (a red M star).

In the upper right of the H-R diagram, well above the cool end of the main sequence, are stars big, bright, and red. Those roughly a hundred times brighter than the Sun are called red giants. Mira, a star in the constellation Cetus whose light varies, and Gamma Crucis, the star that tops the Southern Cross, are examples (both spectral type M). Related to red giants are yellow giants (such as G-type Capella) and orange giants (such as K-type Arcturus and Aldebaran). Giants are rare, accounting for less than 1 percent of stars, but because of their brilliance they enhance the night sky's beauty. Even brighter and rarer than giants are supergiants, shining at the very top of the H-R diagram. Rare though they are, these stars shine so brilliantly that five of the twenty brightest stars belong to the class: the red supergiants Antares and Betelgeuse, the blue supergiant Rigel, the white supergiant Deneb, and the yellow-white supergiant Canopus. Giants and supergiants are nearing the ends of their nuclear-burning lives.

Finally, huddling at the bottom of the H-R diagram, below the main sequence, are the white dwarfs, stars small and faint. Like main-sequence stars, and contrary to their name, white dwarfs actually come in all colors from blue to red. White dwarfs are dying cinders that make up about 5 percent of the Galaxy's stars. They no longer burn nuclear fuel.

ABOVE: The H-R diagram, which plots stellar luminosity against stellar color, reveals main-sequence stars, giants and supergiants;

MAIN-SEQUENCE STARS

Most stars spend most of their lives on the main sequence, just as most people spend most of their lives as adults. This is why main-sequence stars are so common. Indeed, the main sequence is the stellar equivalent of adulthood: the star's youthful volatility has largely vanished, while death is still a long way away. Yet main-sequence stars encompass such a diverse mix of stars—from bright blue powerhouses to feeble red embers—that without the H-R diagram, which connects these stars, they might seem completely unrelated.

The fundamental trait that all main-sequence stars share is this: every one, bright or faint, generates energy the same way, by fusing hydrogen into helium at its core. A star not on the main sequence, such as a red giant or white dwarf, either does not burn hydrogen or does not do it at its core.

Main-sequence diversity stems from one simple quantity, mass. The more mass a main-sequence star has, the hotter its core and the faster it burns hydrogen, so the hotter, bluer, and brighter the star shines. After its birth, a star therefore settles onto the main sequence at a luminosity and temperature that depend on its mass. High-mass main-sequence stars are blue, less massive main-sequence stars are white or yellow, and the least massive main-sequence stars are orange or red. The main sequence, then, is really a mass sequence.

The most massive stars, born with between 16 and 100 solar masses, become hot, blue main-sequence stars of spectral type O. They outshine the Sun thousands of times over, but they pay a price for their brilliance. Though born with more fuel than other stars, they burn it rapidly and die fast. O-type main-sequence stars are so rare that none resides near the Sun. They are rare because few are born and all die within just 30 million years—less than 1 percent of the Sun's age.

The next spectral type, B, contributes greater numbers of blue stars to the Galaxy. B-type main-sequence stars range from 2.5 to 16 solar masses. One such star, Regulus, marks the heart of Leo the Lion.

Another, the southern star Achernar, is the brightest in the constellation Eridanus. Like O-type stars, B stars are short-lived. Regulus, for example, will die within just 100 million years of its birth—far less time than the 4.6 billion years intelligent life took to evolve on Earth.

The white main-sequence stars of spectral type A shine brightly, too, and in such large numbers that many speckle the sky: Sirius, the brightest nighttime star, Vega in Lyra, Altair in Aquila, and Fomalhaut in Piscis Austrinus. These stars have between 1.6 and 2.5 solar masses and live roughly a billion years—probably too short a time for intelligent life to evolve, but perhaps sufficient for primitive life. Because these stars exceed the Sun's luminosity, a planet with a mild climate would have to lie much farther from one of these stars than the Earth does from the Sun. For example, if Vega replaced the Sun, the Earth would have to sit between the orbits of Jupiter and Saturn in order to stay sufficiently cool for life.

THE MAIN SEQUENCE IS the stellar equivalent of adulthood

F-type main-sequence stars, which are yellow-white, outweigh the Sun only moderately, having between 1.1 and 1.6 solar masses. Procyon is a prominent F-type star that has evolved off the main sequence slightly; it is the nearest F star to Earth. The cooler of the F stars live longer than the Sun is old, so some of their planets could have intelligent life.

sunspots, so stars can have starspots

The most likely abodes of intelligent life, think many astronomers, are planets circling yellow main-sequence stars of spectral type G, because the Sun itself is such a star. These stars, ranging from 0.9 to 1.1 solar masses, burn their fuel fast enough to produce copious light and warmth, which life needs; but slowly enough that they survive for billions of years, giving intelligence time to evolve. G-type main-sequence stars account for about 4 percent of stars, so our Galaxy has billions of these suns. Nearby examples include Alpha Centauri A, the brightest member of the nearest star system to the Sun, 4.4 light-years away, and Tau Ceti, the closest single G-type star to the Sun, 12 light-years away.

Sunlike stars exhibit the same phenomena as the Sun. Just as the Sun has sunspots, so stars can have starspots, sometimes so many that the star actually fades when spots rotate into view. And just as the number of sunspots rises and falls every eleven years, so other solar-type stars have spot cycles of various durations. Spots and flares arise from a star's magnetic field, which the star's rotation intensifies. Fast-spinning stars therefore have stronger magnetic fields and greater spot and flare activity than slow-spinning stars like the Sun, which rotates once a month. Billions of years ago, the Sun may have spun faster, because young Sunlike stars often do. At that time, it may therefore have had more spots and flares, bombarding Earth and its inhabitants with greater solar activity that may have helped or hindered the evolution of life. Many Sunlike stars also go through long periods like the Maunder minimum, the time from 1645 to 1715 when few sunspots appeared and the Earth cooled, probably because the Sun dimmed. Stars now in a Maunder minimum must be subjecting any life-bearing planets to frigid climates.

The orange K-type main-sequence stars, some-times called orange dwarfs, are modest suns, having

and all orange dwarfs live long enough that intelligent life has time to arise. Like their G-type brethren, these stars can exhibit spots and flares. K-type main-sequence stars are more common than G stars, consti-tuting 9 percent of stars. Among their ranks are nearby stars such as Alpha Centauri B, the second brightest of the three stars making up this nearest neighbor of ours; Epsilon Eridani, a young star 10.5 light-years from Earth; and Epsilon Indi, an old star 12 light-years away.

In sheer numbers, the red main-sequence stars of spectral type M swamp their brighter peers. These stars outnumber all other types put together, account-ing for 80 percent of the stellar census. They are faint, cool, and small, so they are usually called red dwarfs. Because they are so common, most of the Sun's neigh-bors belong to this meek stellar class—stars like Prox-ima Centauri, the third and faintest member of this nearby triple star system; Barnard's Star, 5.9 light-years away; Wolf 359, which is 7.8 light-years away; Lalande 21185, lying 8.3 light-years away; and a host of others with equally odd names. Despite the prevalence of red dwarfs, not a single one is visible to the naked eye. The brightest is the obscure Lacaille 8760 in the constellation Microscopium, 13 light-years from Earth, a star that requires binoculars to see.

Red dwarfs live long, because they burn their fuel frugally. The dimmest will endure for trillions of years, hundreds of times greater than the present age of the universe. Despite their longevity, red dwarfs may not be good places to find life. A red dwarf glows so faintly that a planet would have to huddle close by to stay warm. A planet around a red dwarf emitting a hundredth of the Sun's energy would need an orbit much smaller than Mercury's in order to have earthly temperatures, and planets may be unable to form so near a star. Furthermore, any such planet would feel strong tides from the star, which would force one side of the planet to face the star and the other to face away. The planet's bright side might then become too hot and the dark side too cold.

For all their dimness, red dwarfs can be cantankerous. Some brighten suddenly, launching huge flares that can outshine the rest of the star. The prototype red dwarf flare star is Luyten 726-8, also called UV Ceti, the fainter star in a double red dwarf system just 8.7 light-years from Earth. These onslaughts of radiation might pose further problems for any life on a planet circling a red dwarf.

orbiting a red dwarf star called Gliese 229. Gliese 229 B is so cool that its atmosphere has methane—something never seen in a main-sequence star, whose heat would tear the molecule apart.

STELLAR EVOLUTION

No main-sequence star lives forever, because each fights its own weight. Gravity first brought the star into existence, compressing and heating it until it shone, but ever since, gravity has been a burden, trying to force the star to collapse further. During the star's life, its hot gas pushes outward, and this gas pressure holds the star up against the inward pull of gravity. In addition, the most luminous stars shine so brightly that the outward pressure of the light flowing out of their cores helps hold the star up. Thus a star balances the inward pull of its weight with an equal and opposite outward pressure.

While on the main sequence, a star generates the necessary energy and outward pressure by converting hydrogen into helium at its core. Sooner or later, though, the core runs out of hydrogen, and the star must move on. Exactly what happens depends on the star's mass. Mass therefore governs not only a star's luminosity and color while the star is on the main sequence but also its fate afterward.

The least massive main-sequence stars exit the least dramatically. After burning hydrogen into helium for over a trillion years, far longer than the universe is old, the smallest red dwarfs—those with masses under 20 percent solar, such as Proxima Centauri—will slowly contract, heat up, and brighten. The red star will turn orange and then yellow, all the time remaining fainter than the Sun. Then the star will cool and fade, turning orange and red again, and eventually dark. If these dead red dwarfs were plotted on the H-R diagram, they would appear at the cool end of the white dwarf sequence; however, no stars have ever entered this state, because the universe is much too young.

NO MAIN-SEQUENCE star lives forever

Red dwarfs range in mass from 8 to 60 percent of the Sun's; they are the least massive of main-sequence stars, but still lesser stars also exist. Called brown dwarfs, these stars fail to achieve even the modest success of red dwarfs, because their centers never get hot enough to sustain nuclear fusion and the stars never join the main sequence. Instead, they shine by gravitational energy. As they age, they simply fade and cool, turning from red to black. Because brown dwarfs are dim, astronomers had a hard time finding them. The first definite brown dwarf was discovered in 1995, just 19 light-years from Earth,

STAR DEATH I: RED GIANTS, PLANETARY NEBULAE, AND WHITE DWARFS

Heftier stars, like the Sun, evolve in a more complicated way, which all stars born with between 0.2 and 8 solar masses follow. This mass range includes a vast array of main-sequence stars: some blue B stars, like Regulus; all white A stars, like Sirius and Vega; all yellow-white F stars; all yellow G stars, like the Sun; all orange K stars, like Epsilon Eridani; and the more massive red M dwarfs, like Lalande 21185.

Take the star of special interest to us: the Sun, whose core now fuses hydrogen into helium. In about 6 billion years, the Sun's core will have turned its hydrogen into helium, but the core is too cool to burn helium, which is harder to fuse. So the star will instead fuse hydrogen in a layer outside the helium-saturated core. The star will brighten and expand. An expanding gas cools, so as the star expands, its surface cools. On the H-R diagram, the star will move upward, because it is brighter, and rightward, because it is cooler, and become an orange and then a red giant hundreds of times brighter than the present Sun. At this time, the Sun will engulf Mercury, and Saturn's frozen moon Titan will become so mild its ice will melt.

Meanwhile, the giant star's helium core starts to stir. The hydrogen-burning shell has been depositing more helium onto the helium core, which has been contracting and getting hotter and denser. Eventually the helium ignites, fusing into carbon and oxygen. The star shrinks and fades, but the surface heats up, and the star becomes a yellow or orange giant. Outside the helium-burning core, the giant continues to burn hydrogen.

As the star's core burns helium, the core fills with carbon and oxygen; when the core exhausts its helium, the helium burns in a shell outside the core, and the star expands, brightens, and reddens, once again becoming a red giant. The Sun will then engulf Venus and possibly the Earth. The star may pulsate, contracting and expanding like a heart, and these pulsations would cause it to brighten and dim periodically. The first pulsating red giant discovered was Mira, which pulsates once every eleven months or so. When brightest, Mira outshines most other stars in its constellation, Cetus; when faintest, Mira can't even be glimpsed by the unaided eye. These pulsations drive mass off the star.

At some point, the red giant completely ejects its remaining atmosphere. Through a small telescope, this cast-off atmosphere looks like a planet's disk, so it is called a planetary nebula, even though it has nothing to do with planets. Famous planetary nebulae include the Ring Nebula, which resembles a multi-colored smoke ring in Lyra, and the Dumbbell Nebula in Vulpecula.

THE SUN will then engulf Venus and possibly the Earth

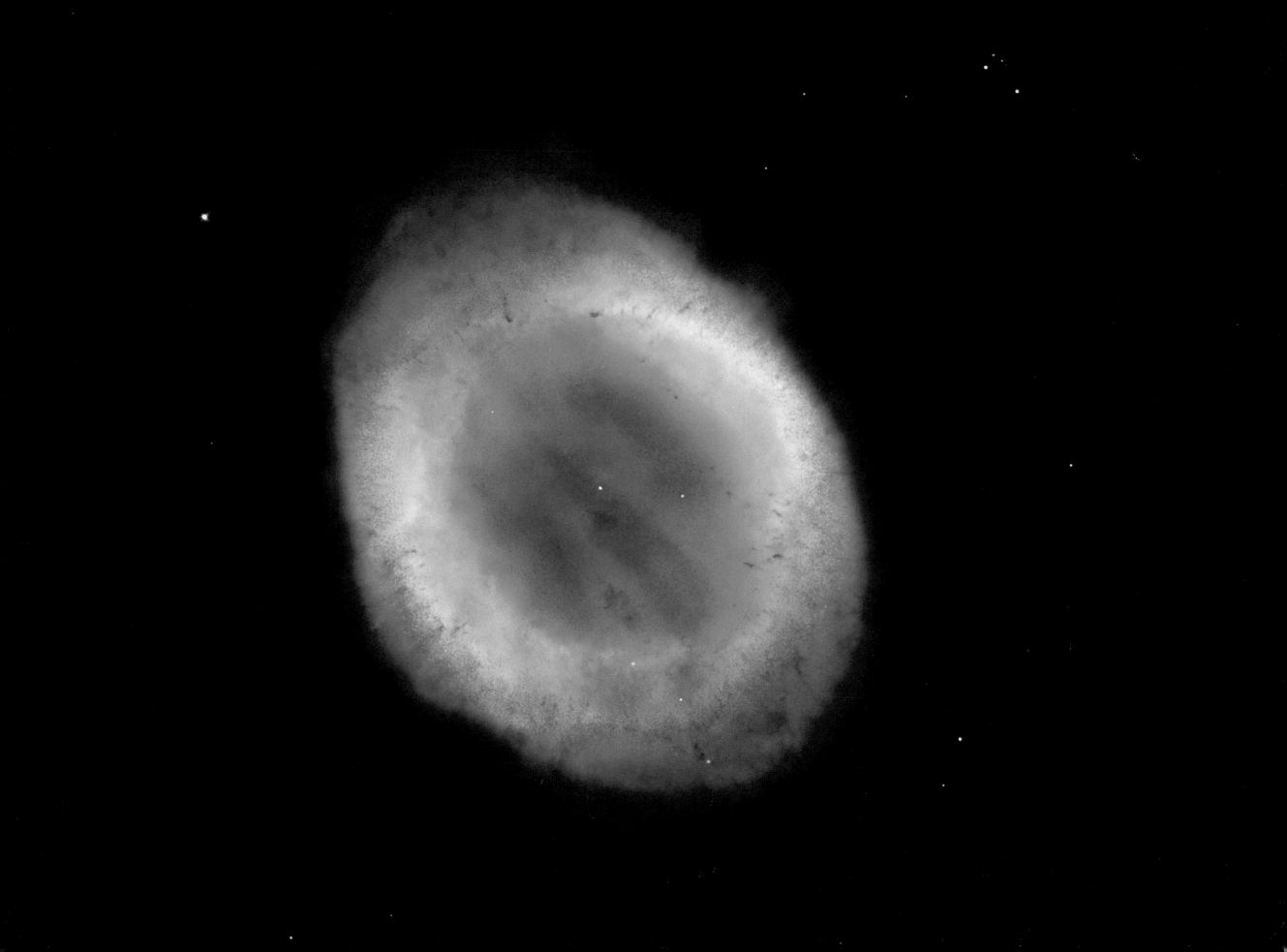

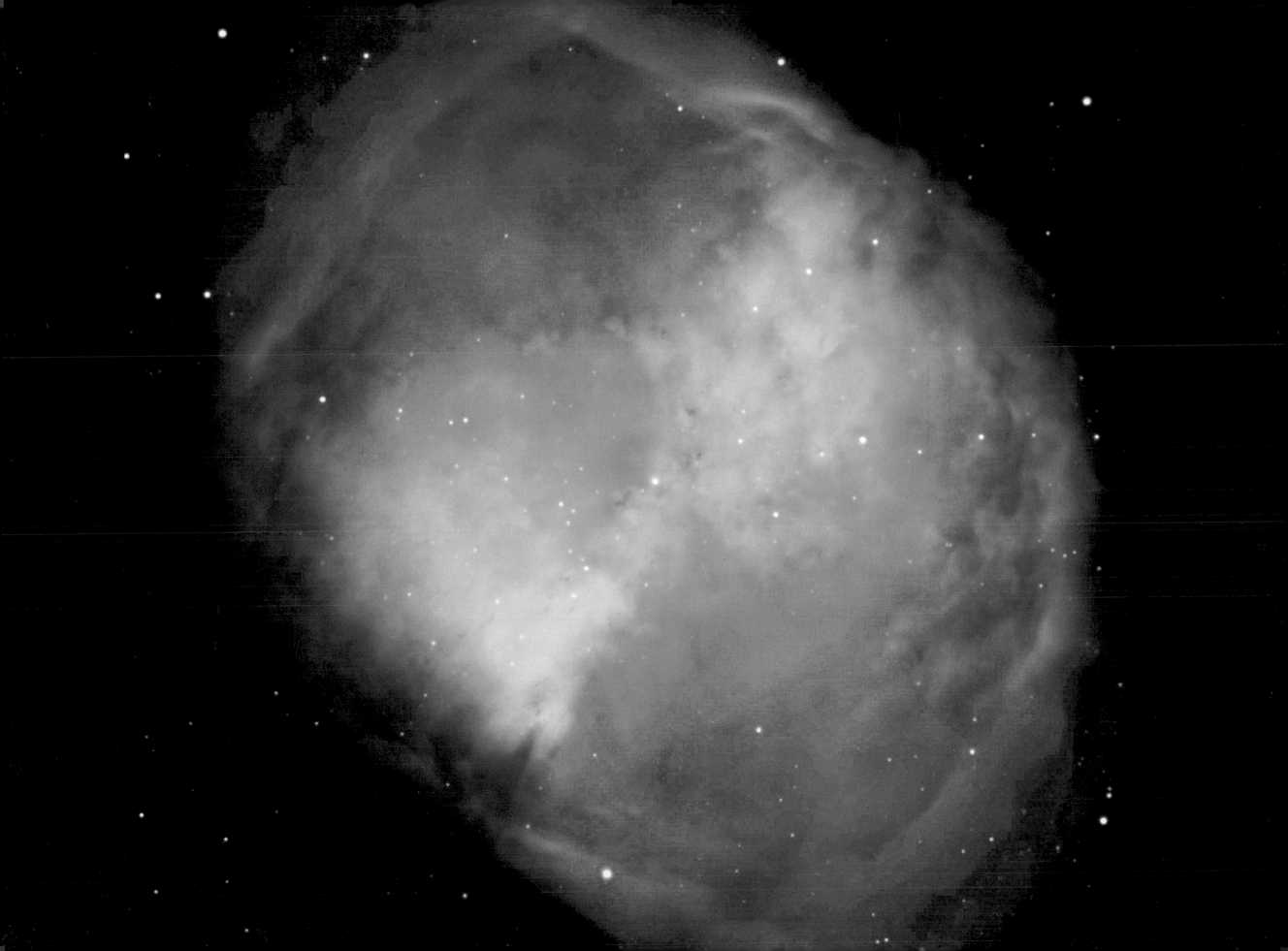

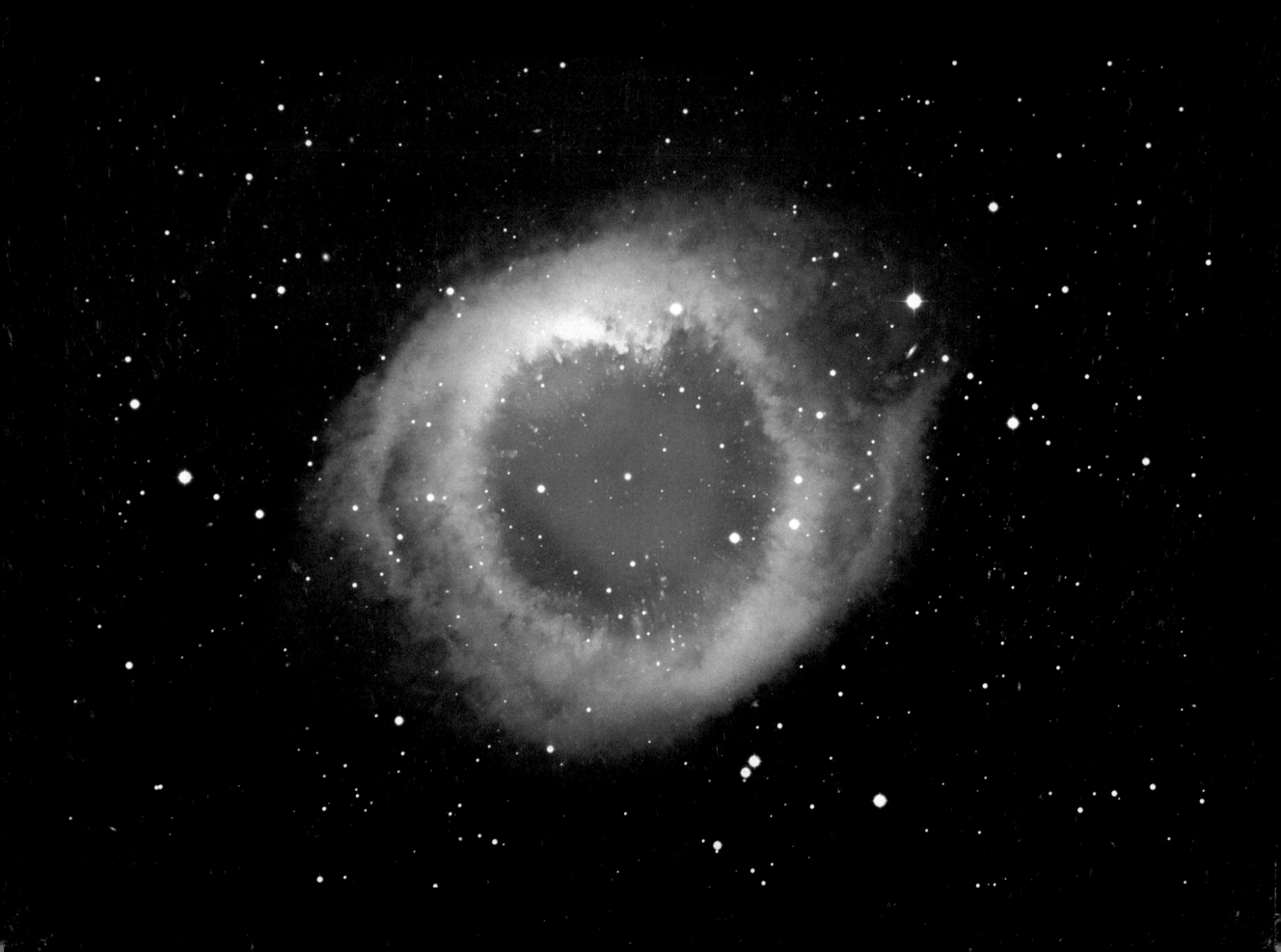

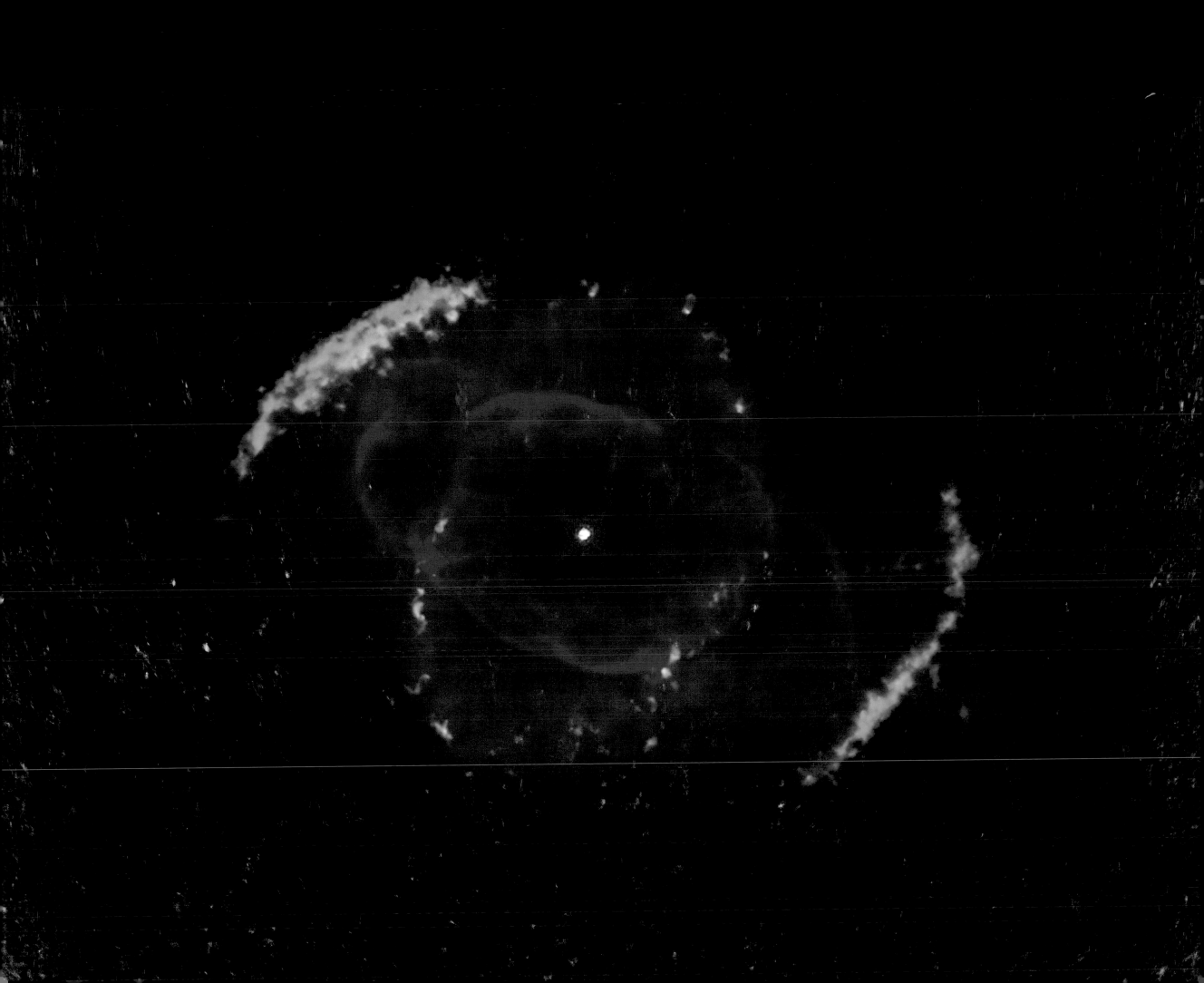

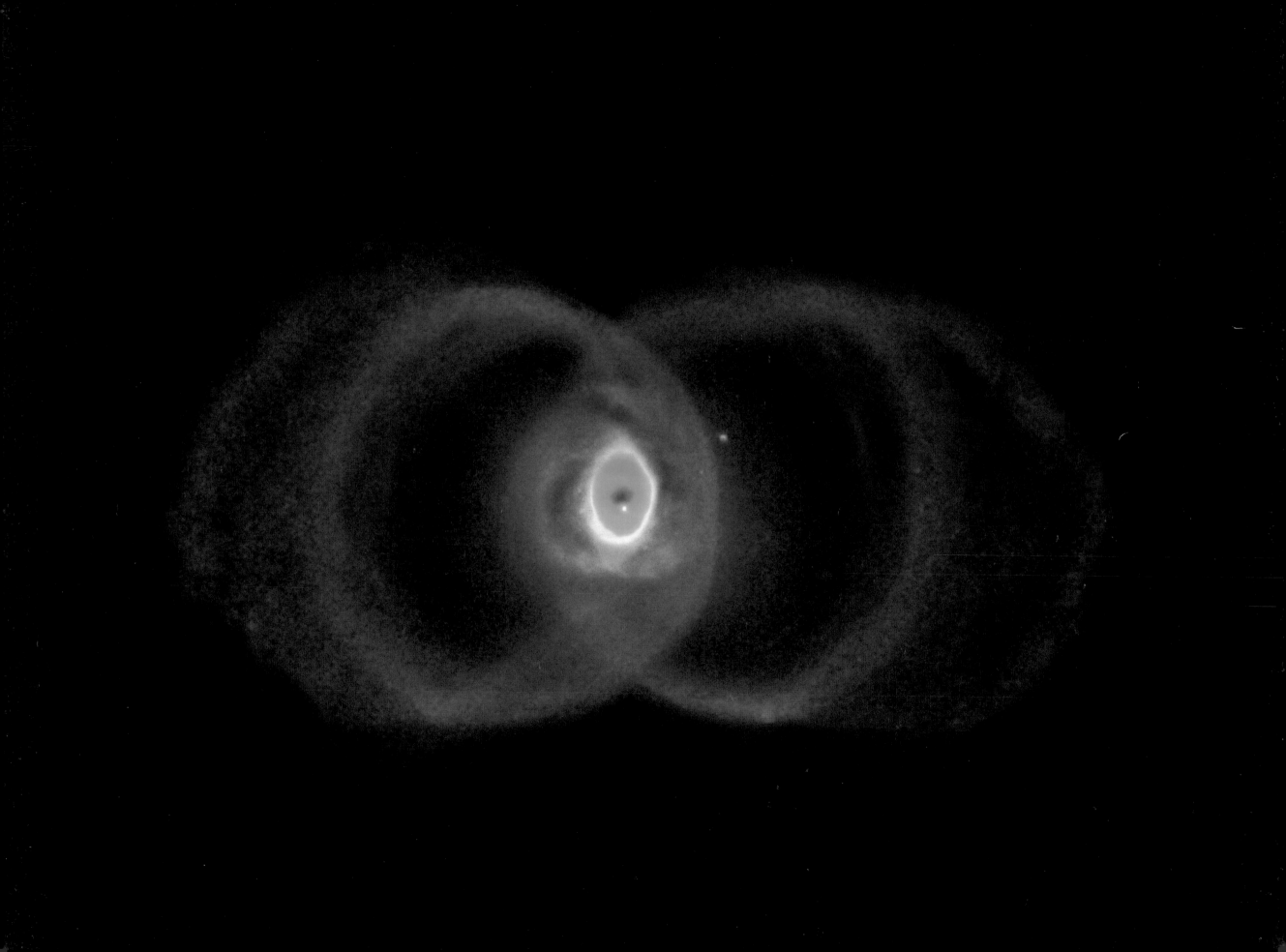

small but hot core emerges, like a butterfly from a cocoon. The core is so hot it radiates ultraviolet light, which sets the planetary nebula's gas aglow and allows observers to see it. Nuclei of planetary nebulae are the hottest stars of all, even hotter than blue main-sequence stars, with temperatures often exceeding 100,000 Kelvin. On the H-R diagram, a planetary nebula nucleus lies well beyond the left edge of the main sequence. It no longer burns fuel; it shines simply because it is hot, and as its heat escapes into space, it cools and fades. The blue star turns white, then yellow, orange, and finally red. Once a big bright red giant, the star is now one of the dimmest lights in the entire Galaxy.

These faint stars are the white dwarfs, which on the H-R diagram lie well beneath the main sequence. A typical white dwarf carries 60 percent as much mass as the Sun, having lost much material while it was a red giant, but is little larger than Earth; therefore, a white dwarf is so dense that a spoonful of its material would weigh many tons. Because so much material is compressed into such a small sphere, the gravity at the surface of a white dwarf is immense: drop a pebble from a height of four feet and it would smash into the white dwarf's surface at four thousand miles per hour. White dwarfs account for one star in twenty, but glow so weakly that all elude the unaided eye. The nearest white dwarfs to Earth orbit two much brighter stars, Sirius and Procyon, 8.6 and 11.4 light-years away.

FACING PAGE: Some planetary nebulae, like the Hourglass Nebula, assume exotic shapes.

STAR DEATH II: SUPER-GIANTS, SUPERNOVAE, NEUTRON STARS, AND BLACK HOLES

The most massive stars, born with over eight times the Sun's mass, live life to the fullest–and die the fiercest. On the main sequence, they are hot blue O or B stars. In order to power their luminous lifestyles, these stars burn fuel at a furious pace. Their first fuel is the same one the Sun burns, hydrogen, which they, like other main-sequence stars, fuse into helium at their cores. When their cores run out, they burn hydrogen outside the core, just as the Sun will. At this stage, however, rather than expand into red giants, as Sunlike stars do, they swell into red *supergiants* that pour out some ten thousand times more light than the Sun.

Red supergiants dwarf mere red giants, shining roughly a hundred times brighter. Two red supergiants, though hundreds of light-years from Earth, rank among the brightest stars in the night sky: Betelgeuse in Orion and Antares in Scorpius. Put either monster in place of the Sun and it would swallow Mercury, Venus, Earth, Mars, and the asteroid belt. The largest red supergiants would engulf Jupiter and Saturn, too.

Supergiants come in other colors, such as blue Rigel in Orion and white Deneb in Cygnus. The most massive stars–those born with over fifty solar masses– never turn red, because they blow off their outer layers so that their surfaces stay hot and blue. A famous example is Eta Carinae, a star so unstable its light fluctuates wildly. Although now invisible to the naked eye, during 1843 Eta Carinae outshone every nighttime star but Sirius.

Supergiants that are yellow often pulsate, their light waxing and waning every few days or weeks as they expand and contract. These are the Cepheids, named for the second such star found, Delta Cephei in the constellation Cepheus. Red giants like Mira also pulsate, but sloppily, with brightness changes and even pulsation periods that vary from cycle to cycle. In contrast, most Cepheids pulsate with the precision of a Swiss watch. The nearest Cepheid is the North Star, Polaris, 430 light-years from Earth; it pulsates every four days. Cepheids shine so brightly they can be seen in other galaxies, and they are especially critical to cosmology since they reveal these galaxies' distances. This is because the longer the Cepheid takes to pulsate, the bigger and brighter it is; thus, the Cepheid's period reveals the star's intrinsic brightness. Comparing the intrinsic brightness with the apparent brightness then yields the distance to the Cepheid and its host galaxy.

NOW THE STAR is in serious trouble

After expanding into a supergiant—red, blue, or yellow—the star ignites its core of helium, as the Sun will, and makes carbon and oxygen. Hydrogen continues to burn outside the core. Then, as the Sun never will, the star's center gets so hot that the carbon begins to burn. It produces neon and magnesium. Then the neon burns, creating oxygen and magnesium. Soon the oxygen burns, generating silicon and sulfur, and finally they ignite to produce iron and other elements of similar mass. The star resembles an onion, each layer fusing a different fuel and churning out energy: an outer layer burns hydrogen, a deeper layer helium, and still deeper layers carbon, neon, and oxygen, all surrounding the silicon-sulfur core that has just sprung into action.

Now the star is in serious trouble. All its life, in order to hold up its tremendous weight, the star has needed to generate a huge outward pressure by burning a huge amount of fuel. The star started by burning hydrogen, a fuel so potent it powered the star for millions of years. When the star switched to helium, it found that the new fuel failed to pack the same punch and so consumed its supply ten times faster. The next fuels, carbon, neon, and oxygen, were even worse. In a final fit of desperation, the star sparked its silicon and sulfur, fuels so feeble they vanished in mere *days*.

The star now has a heart of iron, the most merciless of elements, for iron does not burn. No matter how hard it squeezes or how hot it gets, the star cannot fuse the iron into other elements. One of the brightest stars in the Galaxy—so bright it could be seen millions of light-years away—is about to pay the ultimate price for its extravagant lifestyle.

The core of iron sits there, inert. It cannot provide the outward pressure to support the star's immense weight, so the star's outer layers collapse, bounce against its core, and shoot off into space, making a titanic supernova explosion. If during its brief life the star had retained its outer envelope of hydrogen, then large amounts of hydrogen will appear in the supernova's spectrum, and astronomers will classify it a type II supernova. If instead the star had lost its hydrogen envelope, so that the outer layer is now helium, then astronomers will see this element and classify the explosion a type Ib supernova. And if the star had previously lost both its hydrogen and helium, neither element will appear in the spectrum, and the supernova will be classified type Ic.

Whatever the exact classification, a type Ib, Ic, or II supernova marks the death of a high-mass star, one that shone brightly but briefly and then exploded. Several supernovae occur every century in the Milky Way; however, most cannot be seen from Earth because interstellar gas and dust block their light. Each

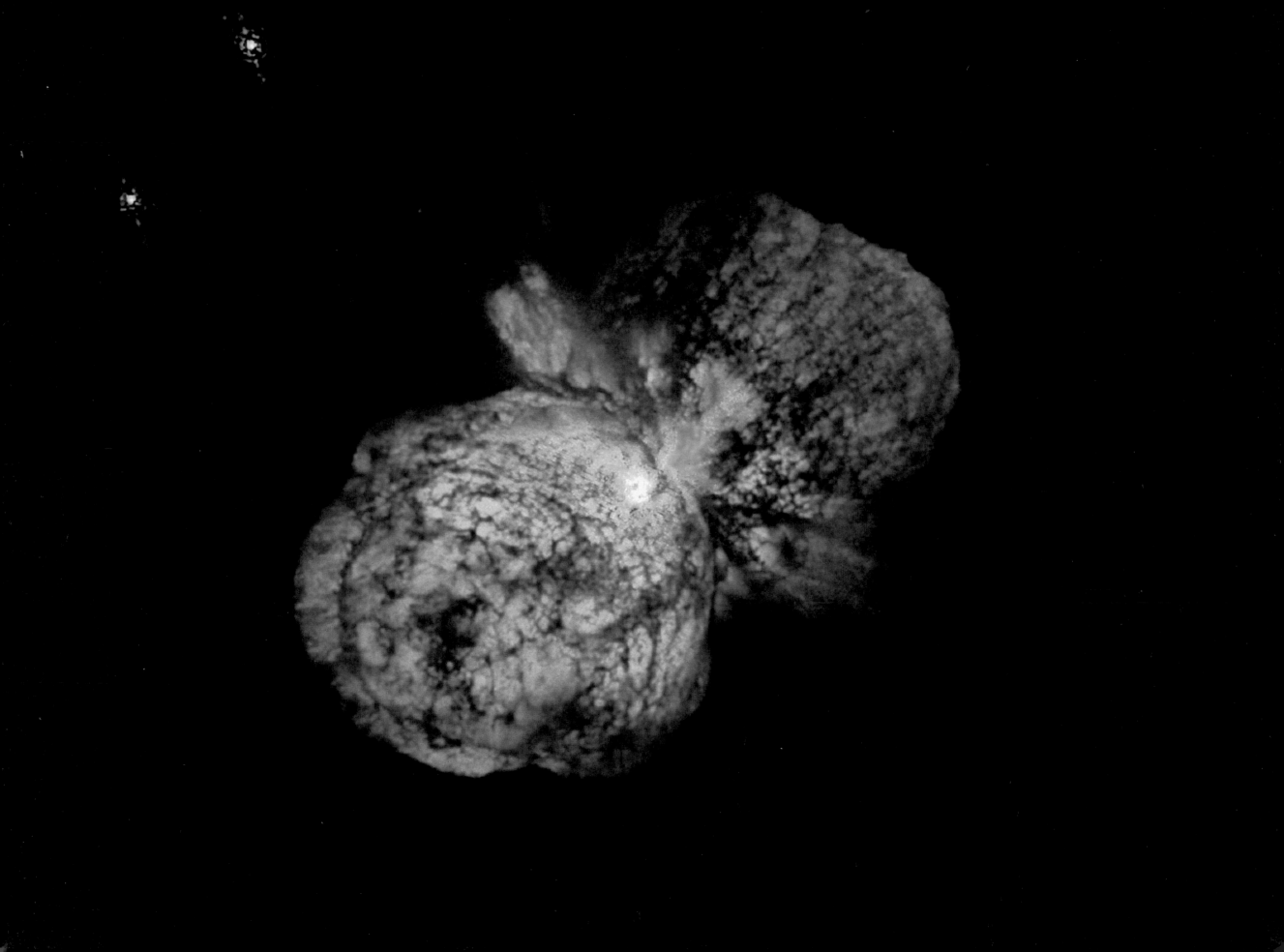

year astronomers observe dozens of supernovae exploding in galaxies beyond the Milky Way, but the last seen within the Galaxy lit the sky four centuries ago, in 1604. Remnants of these explosions litter the Milky Way. One is the Crab Nebula in Taurus, produced by the death of a star our ancestors saw explode in the year 1054. An older supernova remnant lies in Cygnus, the Veil Nebula. It consists mostly of interstellar debris the supernova blast wave has swept up, not of material from the exploded star itself.

A supernova does more than catapult tendrils of gas into the Galaxy. While the outer part of the exploding star was shooting into space at millions of miles per hour, its core was collapsing into an incredibly dense object. Electrons rammed into protons and became neutrons, forming a neutron star. Although just ten miles across, a neutron star typically has 40 percent more mass than the Sun. Because so much material is squeezed into such a tiny object, a spoonful of neutron-star material would weigh billions of tons, and the grav-

FACING PAGE: Eta Carinae erupted in the 1800s.

FOLLOWING PAGES:

Page 94. The Crab Nebula marks the remains of a star that exploded.

Page 95. The Vela supernova remnant consists mostly of interstellar debris swept up by a supernova.

ity at the star's surface is gigantic: drop a pebble from a height of four feet and it would smash into the star's surface at 5 million miles per hour.

A young, fast-spinning neutron star often emits a radio beam that sweeps past Earth every time the star spins, so that astronomers receive a pulse of radio waves with each rotation—just as a spinning lighthouse appears to emit pulses of light. Astronomers discovered the first pulsar in 1967 and since then have cata-

DIVE THROUGH the event horizon and you will never get out

logued over a thousand others. Pulsars spin incredibly fast. The one in the Crab Nebula, born less than a thousand years ago, twirls thirty times a second—nearly a hundred million times faster than the Sun. Pulsars gradually slow down, however, and when their rotation periods exceed a few seconds, they stop radiating and radio astronomers no longer see them. The Galaxy must harbor billions of old neutron stars, the dark, burned-out cinders of stellar spendthrifts.

The most massive stars suffer still worse. If a star is born with more than about forty solar masses, neutrons can't support the collapsed star's weight, and the star shrinks and shrinks until it becomes a black hole—an object so compressed that its gravity won't allow anything, including light, to escape. As far as present physics can say, the entire collapsed star compresses itself into a microscopic point; but space travelers would face danger much farther out. The sphere marking the point of no return—where gravity will not permit light to escape—is called the event horizon. Dive through the event horizon and you will never get out, so neither you nor any radio signals you broadcast will be able to tell the rest of us what you find.

The first likely black hole, Cygnus X-1, was identified in 1971, after astronomers discovered x-rays from a blue star in Cygnus. The blue star circled a dark object that weighed several Suns. A normal star that massive shines brightly, so the dark star was probably a black hole. As the gas from the blue star plunges toward the black hole, but before the gas reaches the event horizon, gravity accelerates it and friction heats it, causing it to emit the x-rays that alerted astronomers. Since then, astronomers have discovered several other black holes around stars in our Galaxy. In addition, a black hole containing 2 to 3 million times more mass than the Sun marks the very heart of the Galaxy, 27,000 light-years from Earth. Other giant galaxies probably have large black holes at their centers, too.

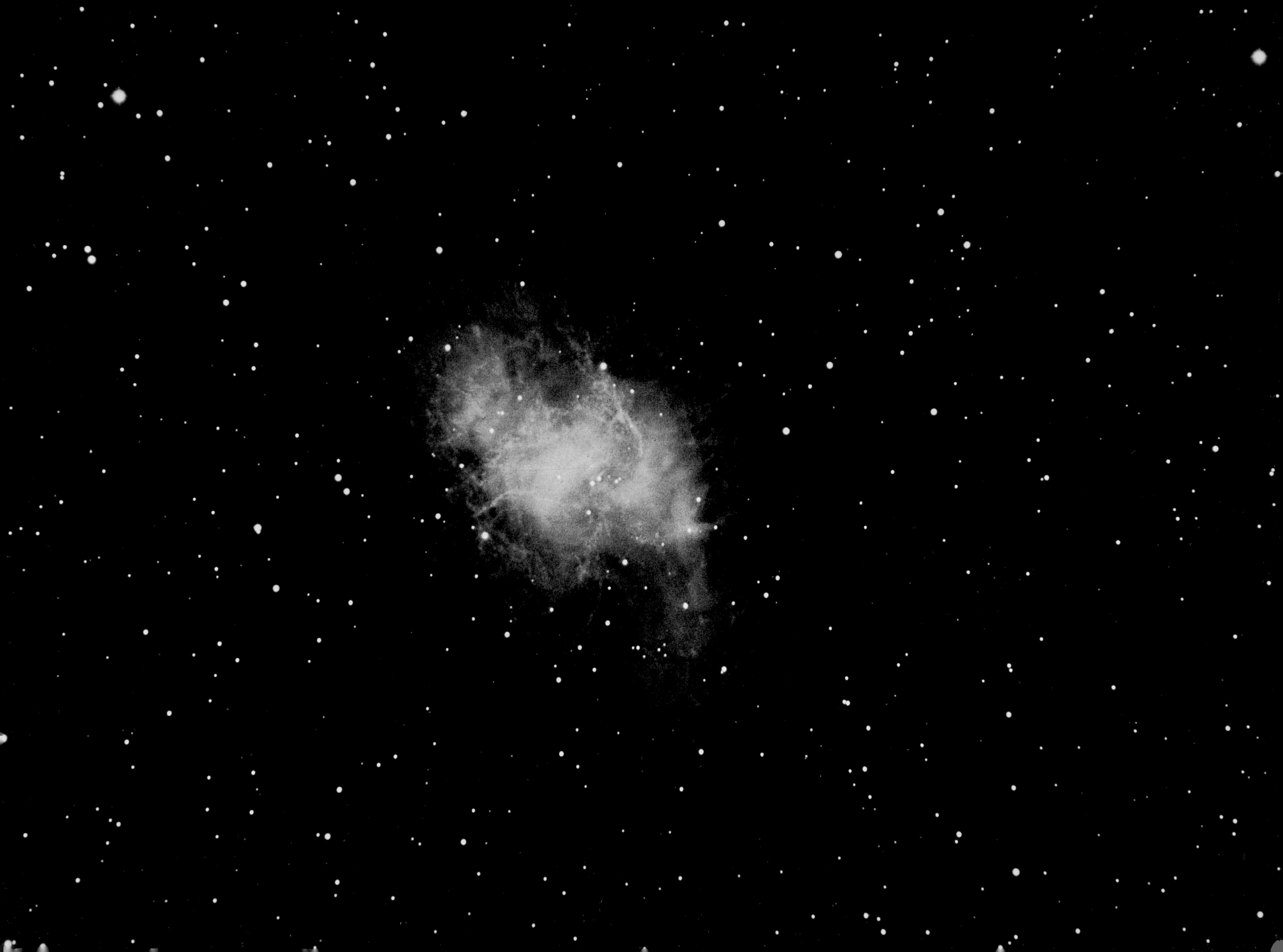

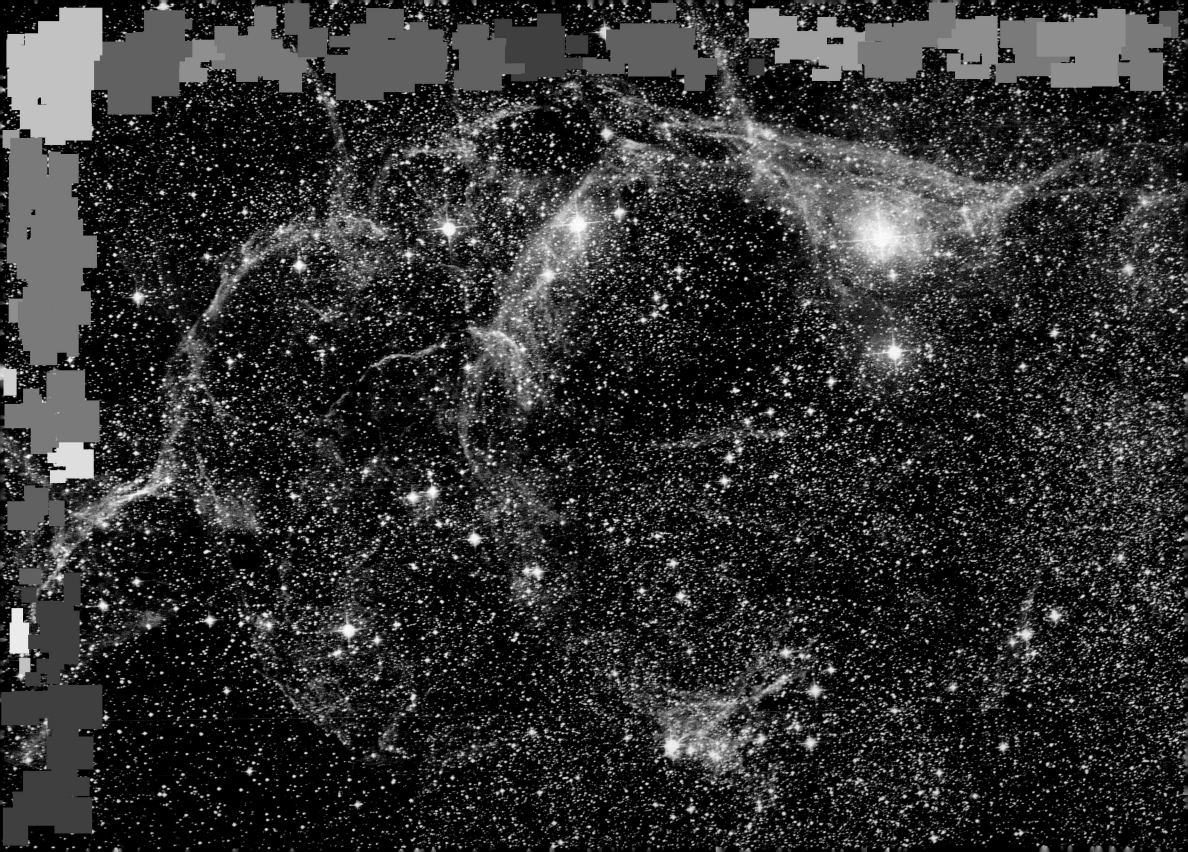

MULTIPLE STARS offer some of the most beautiful sights in the night

star in Cygnus; or Antares, a red supergiant paired with a blue star in Scorpius; or Algeiba, a double yellow-orange star in Leo; or Epsilon Lyrae, which looks double through binoculars and quadruple through a telescope; or Alpha Centauri, a triple with stars yellow, orange, and red. Two of the nearest bright stars, Sirius and Procyon, are double, each luminary attended by a dim white dwarf.

A double star's evolution depends on the distance separating the two stars. If the two lie far apart, they live in peace. Each star can have its own planets, and inhabitants would call the star their planet circles the sun and see the other star as a bright light that at some times during the year appeared in the day and at other times at night. If the two stars instead lie close together, however, one can interfere with the other. Planets could still exist around such a pair, circling both stars, so that any residents would see a double sun in their sky. Some close stellar couples take things to an extreme, forming peanut-shaped contact binaries in which the two stars literally stick together. They were probably born as two separate stars that over millions or billions of years spiraled into contact. Eventually, the two stars will merge into a single spherical star.

shines a bright star that medieval astrologers considered the most dangerous of all, for its light flickers. As if to warn of its malevolent nature, Arabic skygazers long ago named it the Ghoul, or Algol. Algol's light fluctuates because it has two stars that revolve around each other every 2.87 days and periodically eclipse each other. One is a bright blue main-sequence star, and the other is a fainter orange subgiant, a star evolving from the main sequence to the giant stage. When the orange star passes in front of the blue star, Algol fades dramatically.

Algol poses the following paradox. The blue star outweighs the orange one and so should have evolved faster, yet the blue star is still on the main sequence while the orange star is already approaching giant-hood. The resolution to this paradox came from recognizing that the two stars have exchanged matter. At one time the star that is now orange was the more massive, but when it expanded and evolved away from the main sequence, it spilled matter onto its partner, decreasing its own mass and increasing its partner's; therefore, the more evolved star now has less mass than the less evolved star.

Double stars can also explode in ways single stars don't. Each year dozens of novae erupt in our Galaxy, a few of which are seen from Earth. During a typical nova explosion, the brightness shoots up thousands of times, a flare-up in a tempestuous stellar marriage involving a white dwarf star. The white dwarf's companion dumps material onto the white dwarf, and as this material falls and smashes into it, the material gets hotter and denser, eventually igniting the hydrogen and triggering a nova. Unlike a supernova, however, the nova does not destroy the star. Instead, the nova dies down and the white dwarf again begins to accumulate material from its partner, setting the stage for a future nova.

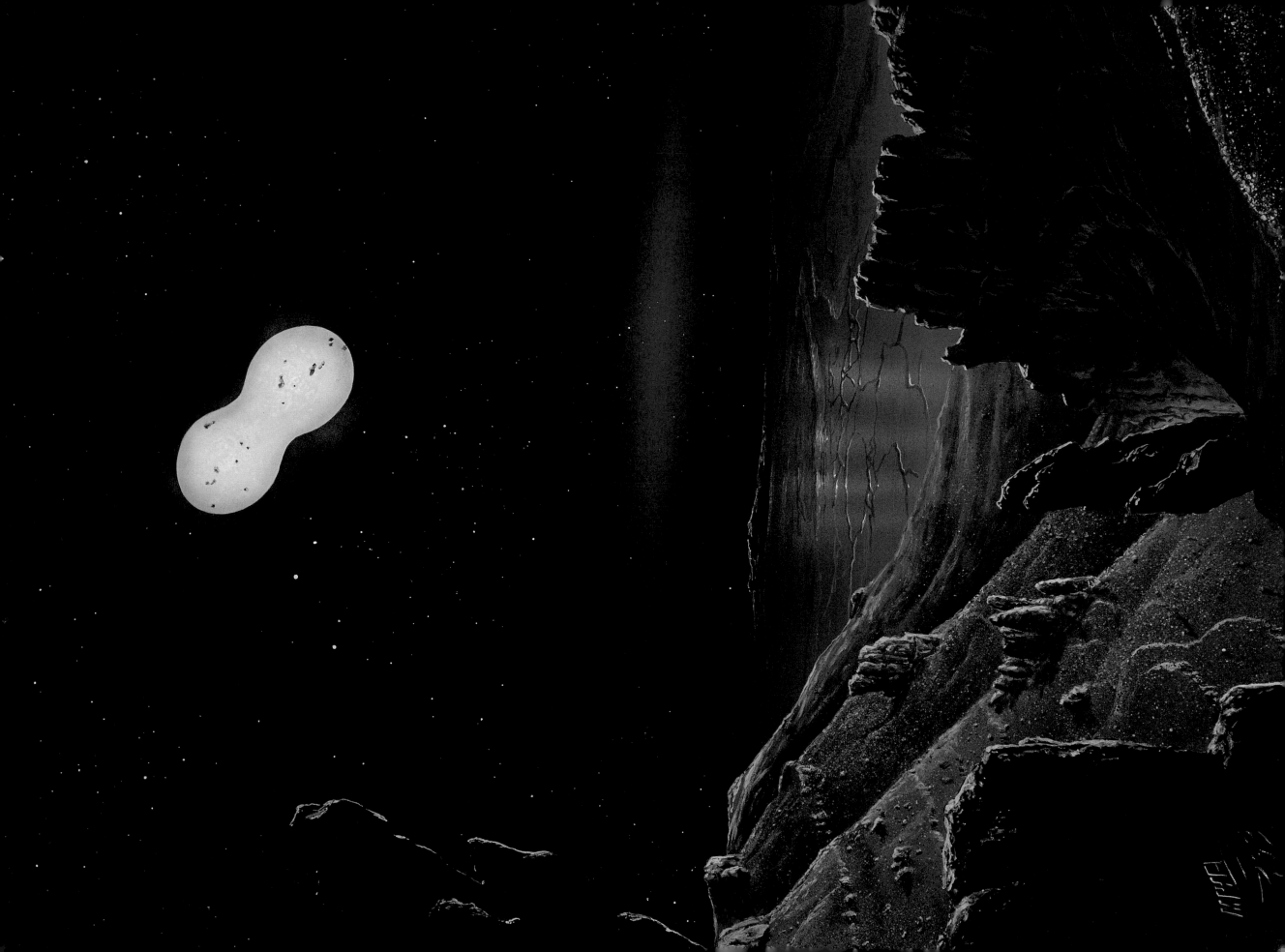

STAR CLUSTERS

Stars can group together still more prolifically. In the constellation Taurus, 400 light-years from Earth, nestles the Pleiades, a beautiful cluster hosting hundreds of young stars. The stars were born together in a nebula like the Orion Nebula and have since traveled with one another a third of the way around the Galaxy. Taurus also houses the Hyades, the nearest star cluster of all, just 150 light-years from Earth. Eventually, though, the Pleiades and the Hyades will disintegrate as the gravitational pull of other stars, giant molecular clouds, and the Galaxy itself tears members from the clusters. Long ago, the Sun may have belonged to a cluster like the Pleiades, but the Sun has since gone its own way, taking the Earth and its other planets with it. Even after a cluster's stars drift apart, they still follow similar paths through the Galaxy and constitute what astronomers call a moving group. The most famous moving group includes the central five stars of the Big Dipper, which were born together 300 million years ago.

Star clusters come in two types. The more common are the open star clusters, like the Pleiades and the Hyades, loose assemblages of roughly a thousand stars. Open clusters usually get torn to shreds within a billion years, so most of them—like the Pleiades and the Hyades—are young. A few open star clusters, however, have managed to survive for 10 billion years. Astronomers can measure a cluster's age by looking at its most massive main-sequence stars. If they are blue, as in the Pleiades, then the cluster must be young, because blue main-sequence stars don't live long; but if they are yellow, then the blue and white stars have died and the cluster must be billions of years old.

STARS CAN group together still more prolifically

Double stars involving white dwarfs also spark far greater fireworks: not mere novae but a special breed of supernova, labeled type Ia, the only supernovae that do not come from high-mass stars. Even though these white dwarf supernovae arise from small stars, they outshine the better-known supernovae that tear big stars apart. A lone white dwarf has nothing to fear, but if a white dwarf has a mate that pours mass onto it, the white dwarf's mass can increase. Danger comes when its mass reaches 1.4 Suns, because then the star's carbon starts to fuse, which generates heat, which causes more carbon to fuse, which generates more heat, which causes still more carbon to fuse, and a runaway nuclear explosion ensues that completely annihilates the white dwarf. Once far dimmer than the Sun, the star spends over a year outshining every other star in the Galaxy. No neutron star or black hole arises. Instead, the entire white dwarf, fried by nuclear fury, splatters into space.

FACING PAGE: A contact binary lights an orbiting planet.

FOLLOWING PAGES:

Page 100. The young stars of the Pleiades cluster travel through space together.

Page 101. Taurus houses both the Pleiades (upper left) and the Hyades, the nearest star cluster; the Hyades appear near Aldebaran (the bright orange star at lower right); however, the star does not belong to the cluster but merely lies in front of it.

Page 102. The Double Cluster, h and Chi Persei, was born just a few million years ago.

Page 103. Few open star clusters survive as long as M67, which is about the

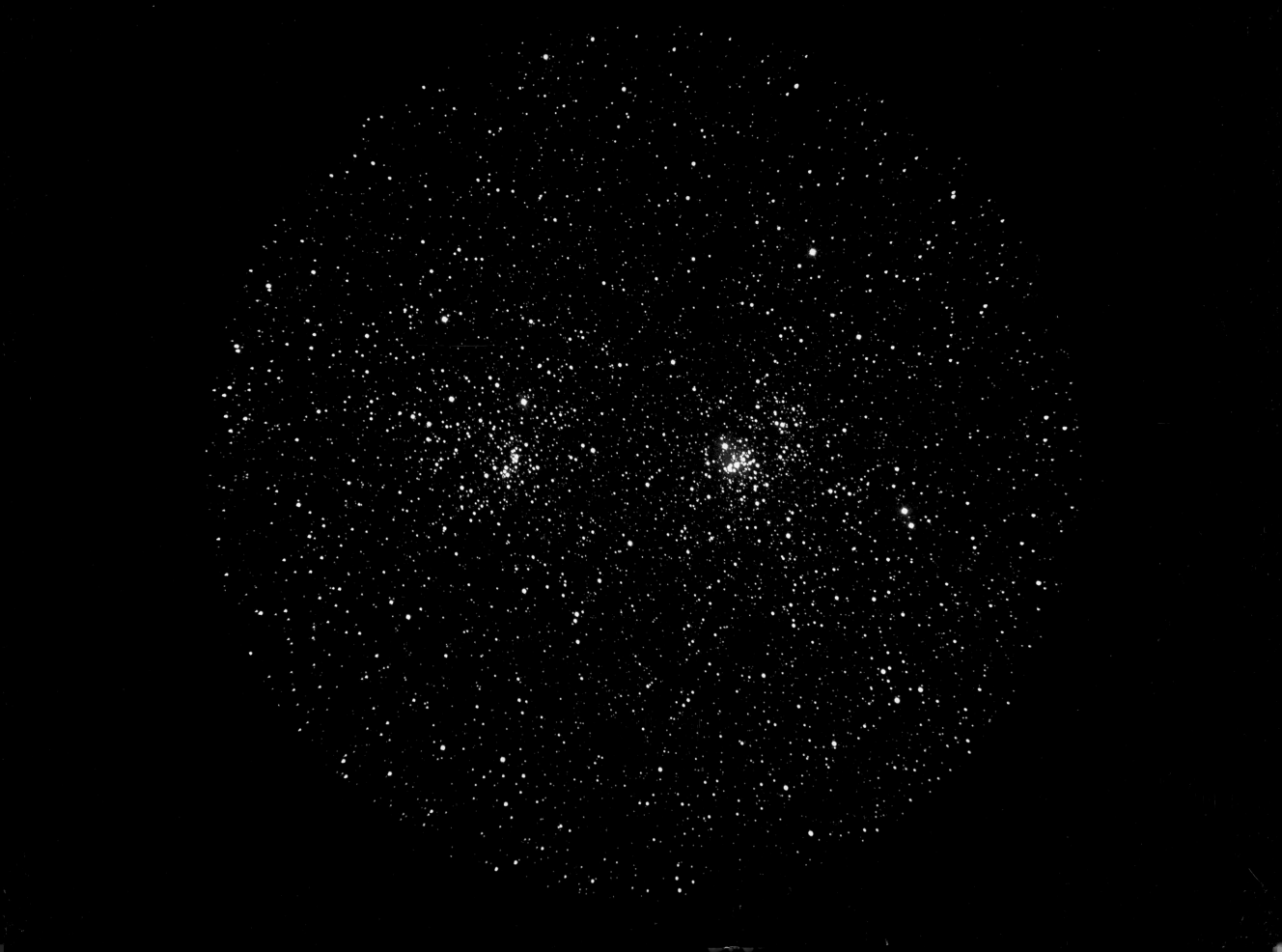

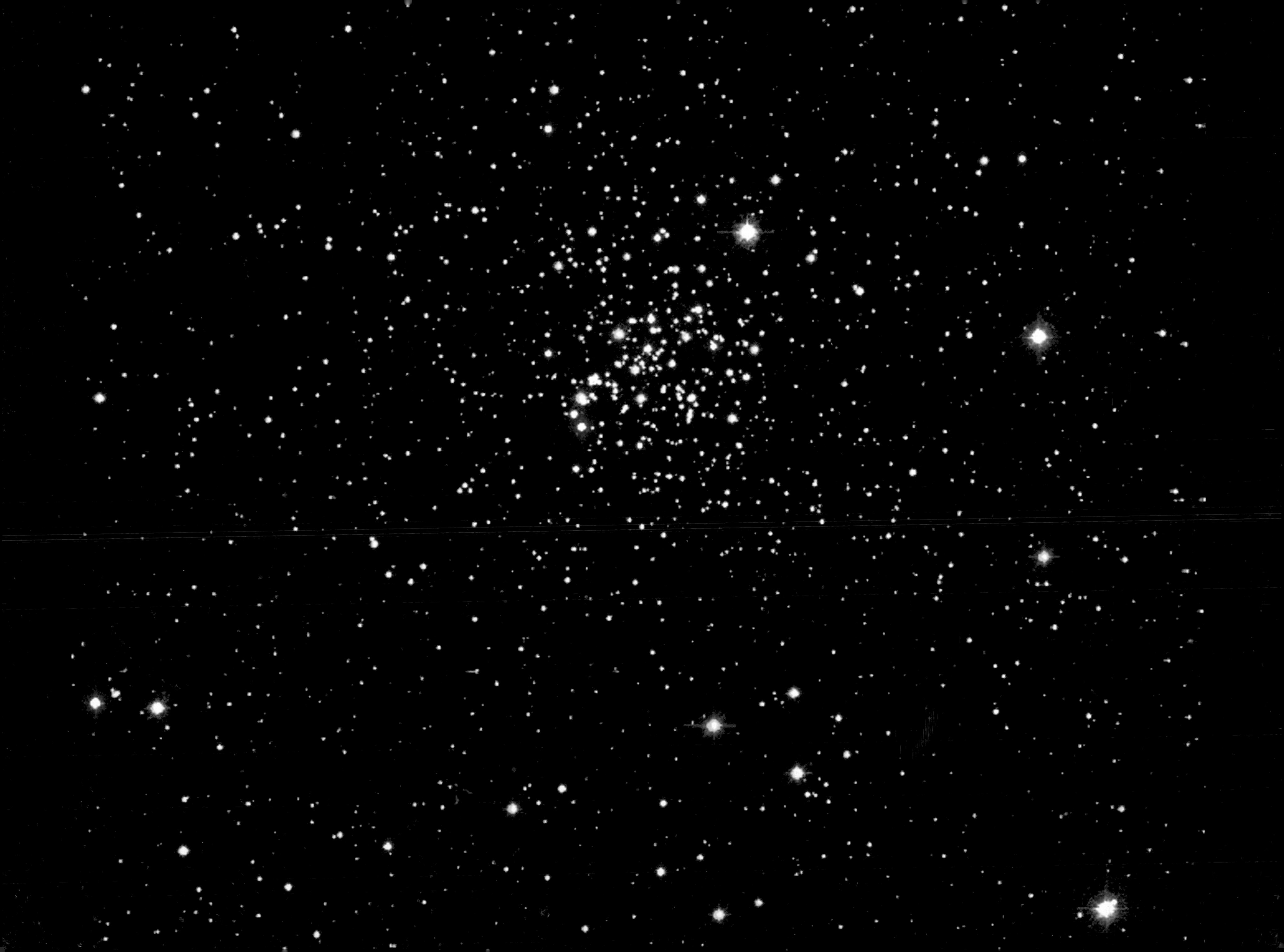

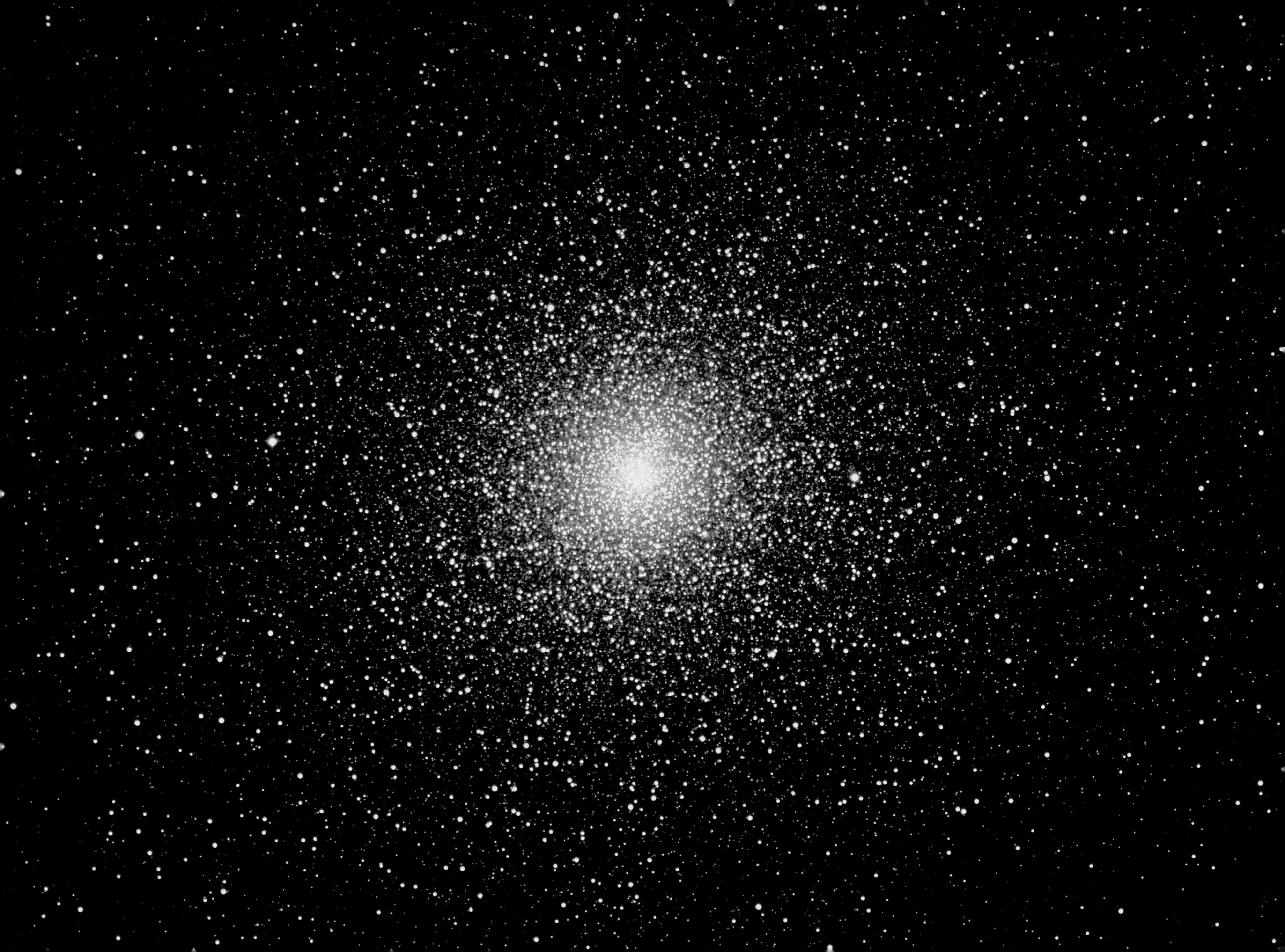

GLOBULAR cluster [pac]ks together hundreds [of] thousands of stars

Open and globular clusters inhabit different parts of the Galaxy. Open clusters, like the Pleiades and the Hyades, reside in the Galaxy's bright disk of stars, as the Sun does; whereas most globulars, like Omega Centauri, belong to the stellar halo, a swarm of old stars surrounding the disk. Most halo stars lie closer to the Galactic center than does the Sun, so most globulars do too. Globulars therefore abound in the direction of Sagittarius, home of the Galactic center. In the same way that a suburbanite sees more city lights when looking toward downtown than away, this concentration of globulars in Sagittarius first alerted astronomers that the Sun does not reside at the Galaxy's center.

died. Long ago, alchemists tried to convert common elements, like copper and lead, into more precious ones, like gold and silver; but true alchemy occurs in the heavens, where the stars transform the most common elements of all, lifeless hydrogen and helium, into the most precious, life-giving elements like carbon, oxygen, and iron.

The short-lived high-mass stars that explode as supernovae work especially hard. During its brief life, a high-mass star desperate to create energy and support its enormous weight manufactures large amounts of oxygen, neon, magnesium, silicon, and other elements that emerge when the star explodes. In addition, the blast of the supernova itself cooks up smaller quantities of still heavier elements, like iron, silver, and gold. More modest stars, which do not explode, generate most of the universe's carbon and nitrogen and eject them when the stars cast off their outer atmospheres in planetary nebulae. Even the dim white dwarfs contribute, for a single type Ia supernova confers upon the Galaxy about 60 percent of a solar mass of pure iron.

This debris gathers in the interstellar clouds that give birth to new stars, so when these stars and their planets arise, they inherit the wealth of their ancestors; therefore, over time, as more and more stars have shed elements into the Galaxy, the Milky Way has grown more and more enriched with these vital, life-giving elements. Without the unnamed ancient stars that preceded the Sun, neither we nor the Earth would exist.

EXTRASOLAR PLANETS AND EXTRATERRESTRIAL LIFE

Of our Galaxy's hundreds of billions of stars, several billion resemble the Sun, being bright, yellow, and long-lived. A planet circling at the right distance from such a star could offer mild temperatures, oceans of water, and perhaps even life. However, astronomers have never detected extraterrestrial life, and only in the 1990s did they begin to discover extrasolar planets. That is because extrasolar planets are extraordinarily difficult to detect: they emit no light of their own; they merely reflect the light of the star they orbit. Furthermore, as viewed from Earth, a planet in another solar system would be swamped by the star's glare.

Fortunately, every planet has mass, whose gravitational pull makes its star wobble back and forth. Since astronomers can see the star, they can see it wobble. These wobbles are minuscule, however, because planets have much less mass than the stars they tug on. By detecting these tiny wobbles, astronomers have snared planets in solar systems beyond our own.

The first extrasolar planets, discovered in 1991 by radio astronomers Alex Wolszczan and Dale Frail, surprised nearly everyone, because the planets circle not a normal, living, shining star like the Sun but instead a pulsar, a dead star emitting deadly radiation. Bearing the prosaic name PSR B1257+12, this pulsar lies over a thousand light-years from Earth in the constellation Virgo. Its planets have roughly the same mass as the Earth, but because they circle such a hostile star, they probably do not harbor life.

In 1995, Swiss astronomers Michel Mayor and Didier Queloz discovered the first extrasolar planet around a star like the Sun—a star called 51 Pegasi, which lies fifty light-years from Earth. The planet is massive, like Jupiter and Saturn, and therefore presumably consists primarily of hydrogen and helium. However, whereas Jupiter and Saturn reside far from

so tightly that it completes an orbit in just four days, over twenty times faster than the Sun's innermost world, Mercury. Since the discovery of 51 Pegasi's planet, astronomers have discovered planets around many other Sunlike stars. None of these planets is likely to support life, since all are giants like Jupiter. Smaller planets, like Earth, could be going around these and other Sunlike stars, providing just the conditions that life requires.

And what are those conditions? Since astronomers presently know only one life-bearing planet—the Earth—no one can say for sure; after all, it's unwise to generalize from a single example. Nevertheless, one prerequisite for life is probably liquid water, an excellent solvent in which chemicals can dissolve and react. Furthermore, unlike most other liquids, water freezes from the top down, so an ocean's surface can freeze while the ocean floor, where life may be struggling to establish itself, can remain bathed in liquid. Terrestrial life probably arose in water, and we still carry this heritage with us, because our bodies are over half water. Water itself is cheap, for its chemical formula, H_2O, means that it consists of the two most abundant reactive elements in the universe, hydrogen and oxygen.

LIFE-BEARING worlds could number in the quintillions

Another requirement for life is energy. On Earth, most life derives energy from the same power that liquifies water, the Sun; but on the ocean floor, some life gets energy from the planet's own heat, through vents and volcanoes; and on Jupiter's moon Europa, life may receive energy from Jovian tides.

The plenitude of stars in the heavens, coupled with the demonstration that planets outside the solar system do exist, means that life-bearing worlds in the observable universe could number in the quintillions.

However, even under ideal conditions—a good planet with oceans of liquid water that orbits a good distance from a good star—no one knows how readily life arises. Some scientists argue that such ideal conditions inevitably lead to life: take water and energy, wait awhile, and life emerges every time. That's what happened on Earth, and may have happened on many other worlds, too. But if it hadn't happened here, we wouldn't be here to contemplate our nonexistence. Unless and until astronomers discover life elsewhere, there is no assurance that any other planet has repeated Earth's remarkable feat.

And life *is* remarkable. Even the most primitive life is so complex that no one can trace the exact steps leading from nonliving material to the simplest life forms. Those steps may be so tortuous and improbable that even a quintillion good planets, with atmospheres and oceans, have all remained lifeless. Right now, no one knows if extraterrestrial life is common or nonexistent—or somewhere in between.

Future years, however, promise to answer this question. A space-based interferometer, which would combine the light from more than one telescope, could yield sharp views of the stars and allow astronomers to see extrasolar planets directly, providing an actual look at small planets like Earth going around bright stars like the Sun. Then astronomers could study these planets' atmospheres, which in turn will reveal much about the planets' natures.

Some of these planets will be nearly airless, like Mercury, unable to support life. Other planets will show carbon dioxide, the predominant gas on Venus and Mars, indicating that the planets at least have atmospheres. Some of the planets with atmospheres will also have water vapor, a sign of oceans. And finally, some of these planets may have atmospheres containing oxygen, a strong sign of life—for life put oxygen into Earth's atmosphere, and oxygen is so reactive that if all earthly life vanished, so would the atmospheric oxygen. The detection of oxygen in another small planet's atmosphere would herald the existence of extraterrestrial life, while the repeated failure to find oxygen around other planets with atmospheres and oceans would warn us that we may be alone.

The number of possible abodes for life in our Galaxy is staggering. After all, hundreds of billions of stars, and perhaps an even greater number of planets,

THE GALAXIES

3

GALAXIES SPECKLE
space the way flowers
dot a spring garden

Galaxies speckle space the way flowers dot a

spring garden. The Earth and Sun inhabit one

of these cosmic blossoms, the Milky Way

Galaxy, a gigantic spiral packed with hun-

dreds of billions of stars. Aside from adorning

the cosmos with splendor, galaxies constitute

the basic building blocks of the universe: they

house the stars that light the dark and form links in long chains

that can span hundreds of millions of light-years. Furthermore, the

largest galaxies, like the Milky Way, have so much mass that their

gravity retains their stars' vital ejected elements—carbon, nitrogen,

oxygen, iron—and recycles them into new stars, new planets, and

possibly new living beings.

All Galaxies Great and Small

The Discovery of Galaxies

Despite their beauty and power, galaxies beyond our own reside at such great distances that they barely embellish the night sky. The unaided eye feasts on thousands of stars but catches less than half a dozen galaxies, all masquerading as faint patches of misty light.

Because of their dimness, galaxies long eluded the astronomical canon. In 1657, English architect Christopher Wren suggested the existence of galaxies outside our own, but in subsequent years many astronomers did not agree. During the 1800s, they discovered numerous spiral-shaped nebulae that were later thought to be newborn solar systems inside the Milky Way. This incorrect characterization actually prompted the investigation that first hinted at their true nature. Early in the twentieth century Percival Lowell, an aficionado of the solar system, asked Vesto Slipher to scrutinize these supposed solar systems. In so doing, Slipher uncovered evidence that they were something far greater. Beginning in 1912, he found that most spiral nebulae raced through space faster than any star in the Galaxy, suggesting they were not attached to the Milky Way but lay in the depths of space beyond. Their remoteness in turn meant they had to be large, for otherwise astronomers could not see them. The spiral nebulae thus seemed to be vast islands of stars that matched the Milky Way itself: "galaxies," after the Greek for "milky ways."

In 1923, Edwin Hubble at Mount Wilson Observatory in California proved this by observing a bright spiral then called the Andromeda Nebula. In Andromeda Hubble detected a Cepheid, a pulsating yellow supergiant star that can indicate an object's distance. By measuring how much time the Cepheid took to brighten and fade, Hubble found its approximate intrinsic brightness; by comparing this with its observed brightness, he deduced its distance and therefore that of Andromeda. The modern value is 2.4 million light-years, so what astronomers now call the Andromeda *Galaxy* shines from well beyond the Milky Way's shores.

Like the stars they lodge, galaxies flaunt diversity. The largest, such as the Milky Way and Andromeda, possess hundreds of billions of stars, and a few titans top a trillion. It is these monsters that most people, including most astronomers, normally envision. Yet giant galaxies actually constitute the minority; for most galaxies hide in dim obscurity, bearing mere millions of stars. Numerically the stars of such a dwarf galaxy compare with those of the Milky Way as the people of Kalamazoo compare with the total population of the world. Galactic luminosities therefore parallel stellar ones: most stars that we see outshine the Sun, but most stars that exist shine feebly, being invisible to the unaided eye; and most galaxies that we see rival the Milky Way's luminosity, but most galaxies that exist muster only a pitiful fraction of our Galaxy's strength.

Nevertheless, every galaxy, bright or faint, large or small, binds its member stars to the homeland through gravity. Surprisingly, this force originates only in part from the masses of the visible stars. Most of it arises instead from a mysterious substance that radiates little or no light—dark matter, which may consist of faint stars or subatomic particles or both. In the Milky Way's case, whereas the light equals only 15 billion times the Sun's luminosity, the dark mass adds up to roughly a trillion times the Sun's mass and gives the Galaxy most of its gravitational strength.

Galaxies differ not only by size, luminosity, and mass but also by shape. In 1926 Hubble published a system for classifying galaxies from their appearance. Among the bright galaxies that he observed, Hubble recognized three types: ellipticals, spirals, and irregulars.

Ellipticals, the least photogenic breed, appear as smooth circles or ovals. A classic example is a giant elliptical galaxy in the Virgo cluster, M87, so named because it is the eighty-seventh entry in French astronomer Charles Messier's catalogue of celestial

objects. An elliptical galaxy like M87 shines intensely at its center and fades fast in all directions. From Earth, astronomers cannot tell an elliptical's true shape, since they view the galaxy from only one perspective. An elliptical galaxy may be spherical, like a tennis ball, oblate, like a pumpkin, or prolate, like a cucumber. Furthermore, a highly elliptical galaxy may actually look spherical, as a pumpkin would seen from above or a cucumber seen end-on. Still, based on their appearance from Earth, Hubble denoted elliptical galaxies from spherical to highly elliptical as E0, E1, E2, E3, E4, E5, E6, and E7. An E0 galaxy looks round, an E1 galaxy's short axis is 9/10 the length of its long axis, an E2 galaxy's short axis is 8/10 the length of its long axis, and so on, and an E7 galaxy is so elliptical that its short axis is 3/10 the length of its long axis. No elliptical galaxies are flatter than E7. M87 is an E1 galaxy.

Although Hubble did not know it when he categorized the ellipticals, their best days have passed them by. Most ellipticals formed all their stars billions of years ago and now consist entirely of old ones. Furthermore, unlike the Milky Way, these galaxies contain little interstellar gas and dust, so they cannot give birth to new stars. Consequently, elliptical galaxies look yellow-orange, because the short-lived white and blue stars have died, and the brightest stars are aging giants, yellow, orange, and red. Some ellipticals, however, do possess younger stars. The nearby small elliptical NGC 205 bears interstellar gas and dust and even a few blue stars, stellar youngsters living in a retirement home.

Elliptical galaxies range from mammoth to minuscule. The largest, rare but spectacular, outdo even elliptical luminaries like M87. They were christened cD galaxies in a classification scheme that tried to supplant Hubble's but otherwise never caught on. If the largest of these behemoths replaced the Milky Way, it would stretch for millions of light-years and completely engulf the Andromeda Galaxy and all other galaxies of the Local Group. cD galaxies look like puffy ellipticals, because their light is more diffuse and dis-

tended. All sit at the centers of galaxy clusters and probably grew fat by gobbling other galaxies.

At the other extreme from the rare cD galaxies huddle hordes of less impressive ellipticals. Many are small ellipticals, such as the aforementioned NGC 205 and its neighbor M32, both satellites of Andromeda. Far fainter are dwarf spheroidals, galactic ghosts unknown when Hubble began to type galaxies. The first dwarf spheroidals, Sculptor and Fornax, were found close to the Milky Way in 1938 but dismissed as exotic star clusters until 1943, when astronomers elevated them into the pantheon of recognized galaxies. Dwarf spheroidals elude detection both because they emit little light and because their stars are spread out from one another. At least eight dwarf spheroidal galaxies orbit the Milky Way, and several others attend Andromeda, so dwarf spheroidals probably outnumber all other galaxies put together.

ELLIPTICAL GALAXIES
range from mammoth
to miniscule

FACING PAGE: Spiral galaxies, like NGC 1232, resemble cosmic pinwheels.

Even within the same Hubble type, spirals display great variety. For example, both M51 and M33 are Sc galaxies, but the former boasts such dramatic arms that it is called a grand-design spiral, whereas M33, though still beautiful, has patchier arms. M51 outshines M33, and indeed luminous spirals tend to be better-looking. Furthermore, M51 has a companion galaxy whose gravity has probably stirred M51 up and intensified the spiral structure. Our own Galaxy is luminous and has two bright companions—the Large and Small Magellanic Clouds—that may have done the same, but no one knows whether our Galaxy is a grand-design spiral.

FACING PAGE: A spiral galaxy hosts stars young and old.

FOLLOWING PAGES:

Page 116. Spiral galaxy NGC 7742 sports an especially bright center.

Page 117. Spiral galaxies have thin disks. This one, NGC 891, resembles the Milky Way.

THE MAJESTIC spirals look like celestial pinwheels

Hubble's second galaxy type, the majestic spirals, look like celestial pinwheels swirling through space. The Milky Way belongs to the spiral class, as does its nearest large neighbor, Andromeda. Most other spirals are large, too. Unlike the typical elliptical, spirals possess interstellar gas and dust and nurture stars young and old. Most of a spiral's stars, and nearly all of its young ones, reside in a flat, pancake-shaped disk. This disk contains the beautiful spiral arms, where interstellar gas and dust get squeezed to give birth to new stars, which color the arms blue. At a spiral galaxy's center sits a bulge of old stars that resembles a small elliptical galaxy, for the bulge is old, yellow-orange, and ellipsoidal. Surrounding the bulge and disk is a diffuse stellar halo bearing ancient stars and globular star clusters, and surrounding the stellar halo and disk is the dark halo, the galaxy's massive reservoir of dark matter. Because of a spiral galaxy's complexity, its appearance depends greatly on the perspective from which astronomers view it. In the most dramatic cases, the galaxy's disk faces us and astronomers see the beautiful spiral pattern. In other cases, however, the disk is edge-on, so astronomers see the disk but not the spiral within. Most spiral galaxies fall between these extremes.

Just as Hubble split elliptical galaxies into classes from E0 to E7, so he divided spiral galaxies into types Sa, Sb, and Sc; later astronomers added type Sd. Although no spiral looks elliptical, the properties of Sa spirals most resemble those of ellipticals, while those of Sc and Sd spirals differ the most. In general, Sa galaxies have tightly wound spiral arms; large bulges relative to their disks; little interstellar gas and dust; little star formation; and few young stars. Sb spirals, such as Andromeda, have more open arms, smaller bulges, and some star formation, while Sc and Sd galaxies have the most open arms, the smallest bulges, the most interstellar gas and dust, and the most intense star formation. Sc galaxies can be especially beautiful, like the stunning Whirlpool Galaxy, M51. Because they

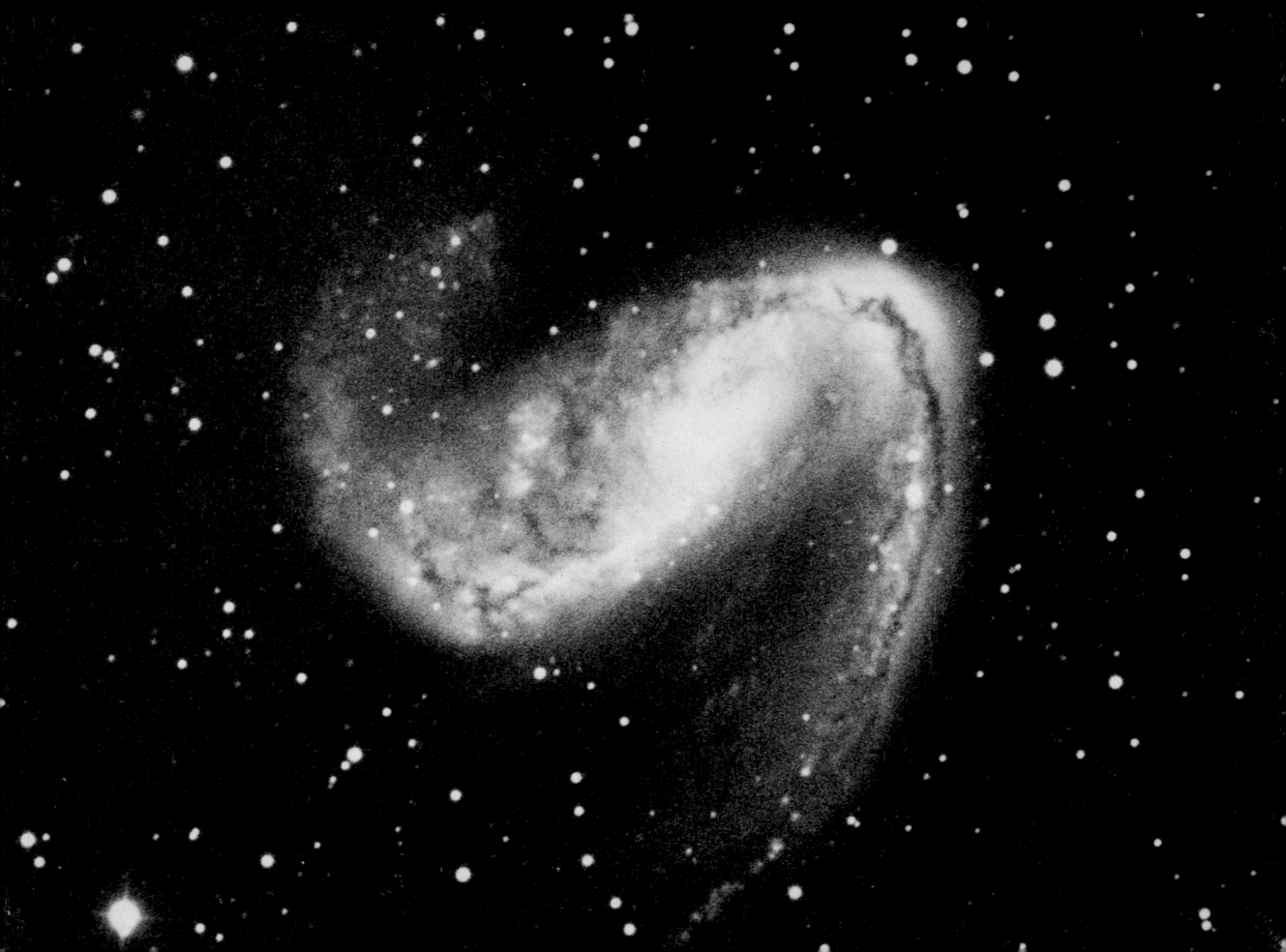

SOME OF the most breathtaking spirals are barred

Some of the most breathtaking spirals are barred. Rather than having a round central bulge, as Andromeda and M51 do, a barred spiral has stretched its bulge into the shape of a cucumber. If spiral arms jut out of the ends of the elongated bulge, the barred spiral can resemble the letter S. The Milky Way itself is probably a barred spiral, but the bar is so subtle that good evidence for it did not materialize until the 1990s.

Whether spiral or elliptical, every galaxy commands its stars through its gravity, so every star orbits the galaxy's center, just as every member of the solar system orbits the Sun. Nevertheless, stars in spiral and giant elliptical galaxies obey their leaders' commands quite differently. In a spiral galaxy, most stars race around the galaxy's center on fairly circular orbits in the galaxy's disk, the way planets circle the Sun. In a giant elliptical, however, stars dart toward and away from the galaxy's center on highly elongated orbits, the way comets do around the Sun. Spirals therefore pos-sess great orbital order, as befits their more beautiful and complex nature, whereas giant ellipticals are orbital jumbles.

In Hubble's system galaxies that fit neither the spiral nor elliptical class are dubbed irregular, like the Large and Small Magellanic Clouds. Irregular galaxies are smaller than most spirals. Although the term may seem cerogatory, irregulars exhibit youthful vigor and vitality, because they abound with gas and spawn lots of new stars, which color them white and blue.

Years after Hubble first classified galaxies, astronomers recognized a hybrid type, the S0 or lentic-ular galaxy, which combines spiral and elliptical traits: like a spiral, it has a disk, but like an elliptical, it has no spiral arms or star formation. S0 galaxies usually inhabit galaxy clusters, where hot gas between the cluster's galaxies may have stripped gas from what were once normal spirals, stunting the formation of new stars and transforming the spirals into S0s.

Because a galaxy consists of countless stars, its complexity exceeds that of any star, just as a nation's complexity exceeds that of any individual within it. Both stars and galaxies emit light, and luminous stars and galaxies are naturally easier to see. For a galaxy, though, light displays an additional property that also influences how easily astronomers can see it: surface brightness, which describes whether the light is strongly concentrated or spread out. All famous galax-ies, like the Milky Way and Andromeda, have high sur-face brightnesses. In contrast, a low-surface-brightness galaxy creeps through the cosmos, an elusive phan-tom. Its stars are spread out from one another so that its light is diluted, like a drop of red dye thrown into the sea. Just as the dye's color fades into the water, so the galaxy's light fades into the sky-making these ghostly galaxies difficult to see. If people were stars and Hawaii and Alaska galaxies, Hawaii would be a high-surface-brightness galaxy and Alaska a low-surface-brightness galaxy, because Alaskans are much more spread out from one another than Hawaiians are.

The first low-surface-brightness galaxies to be discovered, the dwarf spheroidals Sculptor and Fornax, were found in 1938. Over the years, additional dwarf spheroidals were found, and all low-surface-brightness galaxies were thought to be just as puny. But in 1986 astronomers stumbled across a low-surface-brightness monster. Named Malin 1, this spiral galaxy outshines the entire Milky Way, but its light is so diffuse that pho-tographs barely capture its presence. If Malin 1 replaced the nearby Andromeda Galaxy, you still wouldn't be

galaxies, large low-surface-brightness galaxies are usually loners, having no close companions. They constitute an eerie galactic sequence that parallels the normal galaxies most astronomers study.

...ROUPS AND CLUSTERS

Galaxies usually congregate. Most reside in small collec[ti]ons [ca]lled groups, which embrace a few dozen galaxies each. The member galaxies' gravity holds the group together. The Milky Way belongs to the Local Group, whose three dozen or so galaxies also include the Andromeda Galaxy. Groups are sparsely populated, the celestial equivalent of small villages. Their dominant members tend to be stately spirals like the Milky Way and Andromeda, which thrive best in such quiet settings.

...E BEST-STUDIED
...axies inhabit our
...actic backyard,
Local Group

Collections of galaxies greater than groups are called clusters. If groups are celestial villages, clusters are celestial cities, congested with galaxies large and small. A single cluster harbors hundreds or thousands of galaxies, all bound to one another by gravity. Here live many of the giant elliptical galaxies in the universe. The best-known giant elliptical, M87, inhabits the best-known galaxy cluster, Virgo, which lies some 50 million light-years away.

Groups and clusters join to form superclusters. The Local Group, other nearby groups, and the Virgo cluster make up the Local Supercluster, an immense chain of galaxies some 100 million light-years long centered on the Virgo cluster. Beyond the Local Supercluster stretch other superclusters, strung like glistening spiderwebs through the dark. Between superclusters there is mostly nothing, vast voids where few galaxies shine, the enormous undeveloped rural reaches of the cosmos.

Having explored galaxies in general, we now voyage from one specific galaxy to another. We start at home, in the Local Group, sail to other groups nearby and to the more distant Virgo, Fornax, and Coma clusters, and then venture still farther outward, not resting until we sight the most remote outposts in the cosmos, the quasars.

THE LOCAL GROUP

The brightest and best-studied galaxies inhabit our galactic backyard, the Local Group. Most of these galaxies lie within 3 million light-years, so astronomers can scrutinize their stars. Furthermore, in the Local Group, observers can see the faintest of galaxies. The least luminous known galaxy in the universe dwells right here in the Local Group—not because the Local Group really possesses the least luminous galaxy, but because astronomers could never glimpse such a meager entity anywhere else.

FACING PAGE: Like other irregular galaxies, Local Group member Barnard's Galaxy sports gas and young stars.

behind only Andromeda. Gather all the others and you still couldn't construct a galaxy that equaled the Milky Way. So powerful is the Milky Way that it governs ten satellite galaxies and has torn another apart.

Our Galaxy's giant size is no coincidence, because it takes many stars to enrich a galaxy with large amounts of life-giving elements like carbon and iron. Furthermore, when a star explodes, the Galaxy's abundant interstellar material slows the high-speed stellar ejecta before it escapes, and the Galaxy's strong gravity retains that ejecta and pumps it back into more stars. In contrast, small galaxies create only small amounts of heavy elements and have too little gravity to hold on to them. We therefore owe our lives to the Milky Way's large size.

Most of the Milky Way's light emanates from a disk roughly 120,000 light-years across. The Sun resides in this disk, 27,000 light-years from the Galaxy's center. These numbers place the Sun about halfway from the disk's center to the edge, although most books incorrectly claim the Sun is two-thirds of the way out. The main disk is thin, just 2,000 light-years from top to bottom, so in comparison to its diameter, the disk is thinner than a dime. If a knife sliced edgewise through the length of the disk and cut it into two disks, each half as thick as the original, the path the knife took would be called the Galactic plane. The Sun hovers a few dozen light-years north of the Galactic plane, much less than the disk's thickness. Consequently, nearly equal numbers of stars appear on

Although the Local Group has about three dozen members, only two galaxies reign supreme: the Milky Way and Andromeda, which far outweigh all the other galaxies combined. Each superpower commands a retinue of lesser galaxies that revolve around them the way moons circle a giant planet. In fact, most Local Group members orbit either the Milky Way or Andromeda. A few galaxies, though, manage to steer clear of both and maintain a nonaligned status, obeying neither giant but nevertheless bound to the Local Group by gravity. The Local Group therefore splits into three divisions: the Milky Way and its satellite galaxies; Andromeda and its empire; and the independents.

THE MILKY WAY GALAXY

Home of the Sun and the Earth, and once thought by some to be a river running across the sky, the Milky Way Galaxy decorates the night: down dust lanes and spiral arms, illuminated by blue supergiants and littered with remnants of supernovae; through cast-off stellar smoke rings and patches of magenta nebulosity spawning new-born stars and planets; past erupting red dwarfs, quiet yellow suns, and voracious black holes. Our Galaxy offers unique insight into others beyond, for in the Milky Way astronomers can excavate the rich fossil record written in the ages, locations, motions, and chemical compositions of individual stars and thereby reconstruct the birth and evolution of an entire galaxy, much as an archaeologist examines ancient artifacts to decipher the past. At the same time, however, we lie trapped within the Milky Way's confines, enmeshed among interwining star lanes and dust clouds that tangle together like leaves in a tropical forest: our Galaxy is the only galaxy we cannot view from the outside.

The Milky Way is much larger and brighter than most other galaxies. Of the Local Group's many members, the Milky Way ranks an impressive number two,

EVERY STAR bright enough for the naked eye to see belongs to the Milky Way Galaxy

and far, and when we look above or below it we see only the relatively few near the Sun. But whether in this band or not, every star bright enough for the naked eye to see belongs to the Milky Way Galaxy.

The disk features the spiral arms, which give our Galaxy the look of a pinwheel to any outsiders. Here lie the giant molecular clouds as well as the bright young stars and glowing red H II regions they generate. The Galaxy's greatest luminaries, stars like Rigel, Betelgeuse, Deneb, and Antares, occupy the arms, because such spendthrifts burn out before getting far from their birthplace. In contrast, as a long-lived star like the Sun travels around the Galaxy, the star weaves in and out of the spiral arms, like an airplane passing in and out of cloud banks. Since long-lived stars far outnumber highly luminous ones, nearly as many stars populate the dark interarm regions as the bright spiral arms; the latter *look* more dramatic only because they boast the luminous stars that light them.

Celestial spiral structure was first seen from afar, in the "nebula" M51 in 1845, before astronomers even recognized it as another galaxy. Subsequently, astronomers discovered other spirals and speculated that the Milky Way was one as well, but because a stellar thicket entangles us, a hundred years elapsed before astronomers proved it. In 1951, a team led by William Morgan of Yerkes Observatory in Wisconsin mapped the positions of nearby H II regions, such as the Orion Nebula, and found that they ran along two parallel paths. One, named the Orion arm, wound through the Sun and housed the Orion Nebula and the bright stars of that constellation. The other spiral arm, farther from the Galactic center, snaked through Perseus and Cassiopeia and was named the Perseus arm. Two years later, this work revealed a third spiral arm, closer to the Galactic center, the Sagittarius arm. Astronomers have since confirmed the Sagittarius and Perseus arms; the Orion arm may merely be a protrusion from another arm, so astronomers sometimes call it the Orion spur instead.

The Milky Way's structure, and the Sun's location within it, dictate what observers see season by season. During winter and summer, stars abound, because the Milky Way's disk rides high in the sky. The Sun resides near the inner edge of the Orion arm or spur, so when we look *away* from the Galactic center, we see most of the Orion arm's splendors. This we do every winter in the northern hemisphere (summer in the southern), so stargazers find this season the most spectacular. We see bright nearby stars, like Sirius, Procyon, Capella, and Aldebaran, as well as distant ones like Betelgeuse and Rigel; open star clusters like the Hyades and the Pleiades; and nebulae like the Orion Nebula and the Horsehead Nebula. In addition, through the gas and dust choking the Orion arm, we discern denizens of the distant Perseus arm, such as the Double Cluster, h and Chi Persei.

In northern summer (southern winter), we look the opposite way, toward the Galactic center and away from the heart of the Orion arm. We therefore see somewhat fewer bright stars, those between the Sun and the Orion arm's inner edge, stars like the eight that form the lovely Sagittarius teapot and most of the stars outlining Scorpius, whose heart is lit fierce red by Antares. However, because we look toward the Galaxy's center, in Sagittarius, "the Milky Way"—that band of glowing starlight—is thicker and brighter. In that band lie residents of the Sagittarius arm, such as the Eagle, Lagoon, and Trifid Nebulae.

During spring and fall, the night darkens, for we gaze above and below the Milky Way's disk. Now the disk hugs the horizon, and the only stars that look bright are those nearby, such as Arcturus in northern spring and Fomalhaut in northern autumn. Yet these same seasons delight galaxy lovers, for the Milky Way's gas and dust, which block about 20 percent of the extragalactic sky, lie low, allowing observers to see numerous galaxies, such as those in the Virgo and Coma clusters during northern spring and the Fornax cluster during northern fall.

No star's position in the sky endures forever, because every star speeds through the Galaxy. As viewed from north of the Galactic plane, most stars, including the Sun, revolve clockwise around the Galactic center and remain close to the Galactic plane. The Sun dashes through space at half a million miles an hour, taking the Earth and the other planets with it. It completes an orbit every 230 million years, so since its birth the Sun has circled the Galaxy twenty times. As the Sun revolves, it also bobs up and down through the Galactic plane, like a horse on a merry-go-round. The Sun presently lies a few dozen light-years north of this plane and each year climbs 140 million miles farther north, the distance between the Sun and Mars. Some 15 million years from now, the Sun will crest about 250 light-years above the plane and then start to fall back down. After diving through the Galactic plane, 15 mil-lion years later, it will plunge 250 light-years below the plane, and then the disk's gravity will pull it back north again. Not only does the Sun move up and down, but it also winds in and out, its distance from the Galaxy's center varying by 3,000 light-years.

Stellar positions and motions lend clues to the Galaxy's origin and evolution, as do the stars' ages and chemical compositions. To piece together Galactic history, astronomers employ the concept of stellar populations. A stellar population is a galaxy-wide assemblage of stars that share similar ages, locations, orbits, and chemical compositions. At least four stellar populations inhabit the Milky Way: the thin disk, the brightest stellar population, which includes the Sun and most other stars nearby; the thick disk, old stars that climb farther above and below the Galactic plane; the stellar halo, a home for ancient stars that often have wildly elliptical orbits; and the bulge, whose stars wrap around the Galaxy's center. In addition, a halo of dark matter, which may or may not possess stars, envelops all this and extends far beyond the disk. Old stars, such as those in the thick disk and stellar halo, generally possess smaller amounts of heavy elements like oxygen and iron than do the young stars thronging the thin disk, because old stars formed before the Galaxy's stars had a chance to manufacture heavy elements.

At the very heart of our Galaxy shines a peculiar source of radio waves called Sagittarius A* ("ay star"), a disk of matter orbiting a likely black hole that out-weighs the Sun millions of times over. Within just one light-year of Sagittarius A* swarm millions of stars, each caught in the black hole's gravitational grip; by contrast, not a single star lies within this distance of the Sun. The black hole's great gravity forces stars and gas to whirl around it rapidly. When one of these stars sheds matter, the black hole grabs its share and grows still mightier, and occasionally it gobbles entire stars, along with their planets.

FACING PAGE: The Milky Way's center, home of a giant black hole,

Distributing the Galaxy's disk and stellar halo is
he dark halo, which emits little or no light but
stretches perhaps 200,000 light-years from the Galac-
ic center. Invisible though it is, the dark halo harbors
most of the Galaxy's mass, more than all the luminous
stars put together. The dark halo betrays its existence
solely by its gravity, tugging on stars and gas in the
outer disk and making them revolve faster. In addition,
without the dark halo, the Milky Way's satellite galaxies
would escape its grasp.

In the Milky Way's stellar population mix are
clues to the story of its origin and evolution. Some 10
to 15 billion years ago, a ball of hydrogen and helium
gas collapsed, giving birth to stars in the stellar halo.
They had highly elliptical orbits, because the collapse
was rapid, and few heavy elements. But some of the
halo stars exploded, quickly enriching the Galaxy's gas
with oxygen, iron, and other heavy elements. The gas
then settled into a swirling disk, which gave birth to the
stars in the thin disk. Because of the supernovae that
preceded their births, thin-disk stars were born with
healthier supplies of heavy elements. Smaller galaxies
crashed into the Galaxy, providing additional stars to
the stellar halo. The thick disk may have arisen when a
rather large galaxy, perhaps a tenth as big as the Milky
Way, smashed into it, puffing up the thin disk and also
contributing its own stars.

THE MILKY WAY'S SATELLITE GALAXIES

The Milky Way oversees more than just a multitude of
stars; it also rules an empire of at least ten other galaxies. The two brightest are the Large and Small Magellanic Clouds, 160,000 and 190,000 light-years from the
Milky Way's center. These irregular galaxies brim with
gas and churn out young stars that keep the galaxies
bright and vigorous. Because of their proximity and
luminosity, the Magellanic Clouds appear larger and
brighter than any other galaxy but our own. Unfortunately, they lie so far south that most observers at
northern latitudes can't see them. In 1987, light from
the explosion of a young star in the Large Magellanic
Cloud reached Earth, heralding the brightest supernova since 1604, the last time astronomers saw a
Galactic star tear itself apart. Although subservient to
the Milky Way, the Magellanic Clouds are not small
galaxies. Among the Local Group's members, the Large
Magellanic Cloud ranks a respectable number four and
the Small Magellanic Cloud number eight. Furthermore,
the Large Magellanic Cloud flaunts the Local Group's
most luminous H II region, the Tarantula Nebula, where
thousands of newborn stars shine.

The Magellanic Clouds probably move through
the Milky Way's dark halo and therefore get slowed by
the gravity of the small objects—stars or subatomic
particles—that populate it. This so-called dynamical
friction robs the Magellanic Clouds of their orbital
energy, so they will spiral closer and closer to the Milky
Way until, billions of years from now, our Galaxy swallows them and acquires their stars for itself.

At greater distances than the Magellanic Clouds,
eight dim galaxies bearing the names of constellations
swing slowly around the Milky Way. These galaxies are all
dwarf spheroidals, and most manage just a few million
widely separated stars. One dwarf spheroidal, Draco, is
the faintest known galaxy in the cosmos, producing less
light than the brightest single star in the Milky Way.

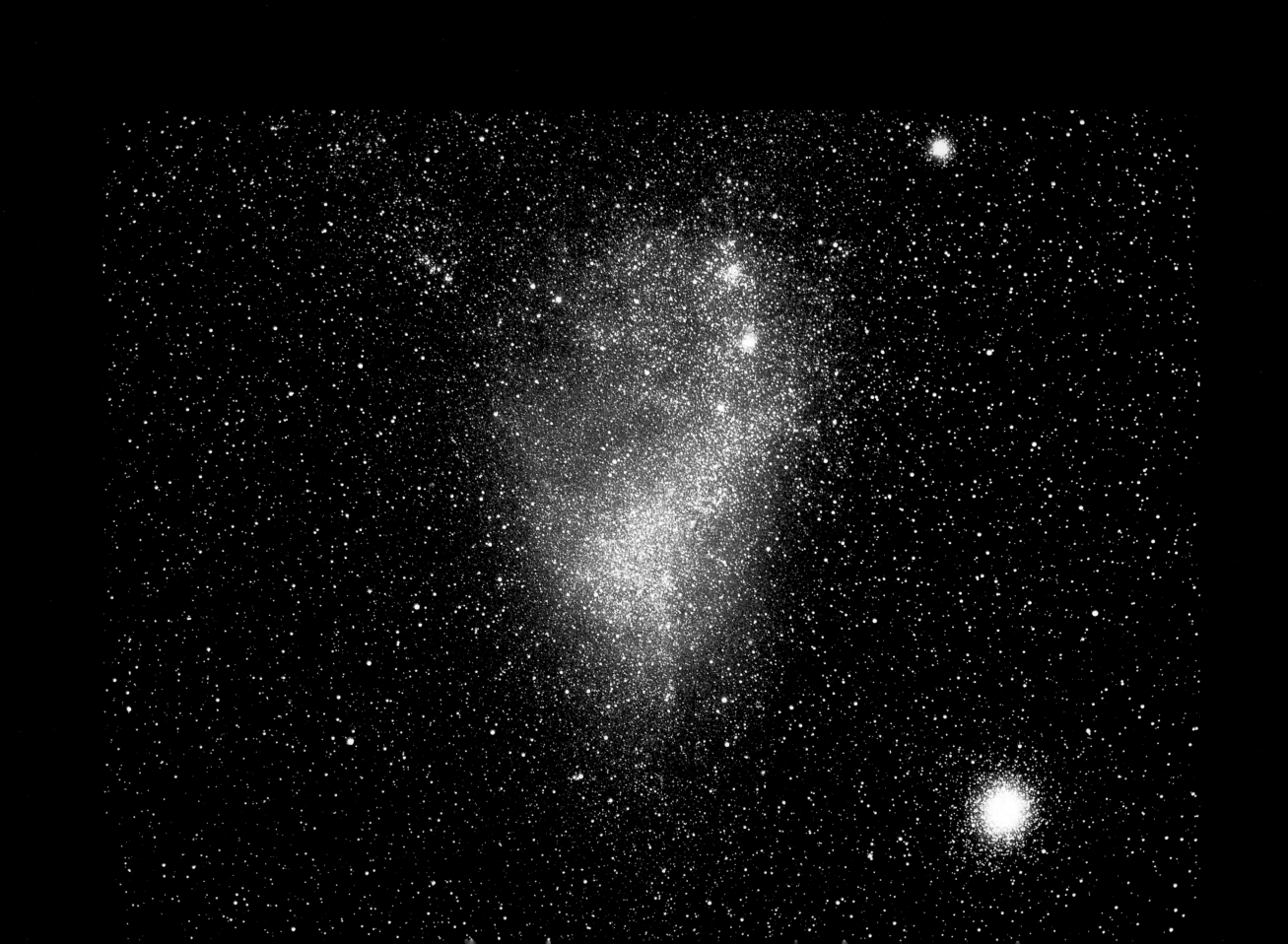

Today, however, Andromeda is the quieter galaxy, manufacturing only about one new star a year, whereas the Milky Way teems with star-breeding molecular gas that spawns ten times as many stars.

LIKE THE MILKY WAY, Andromeda rules an empire of satellite galaxies

During the 1990s, astronomers learned of a link between the dwarf spheroidals' distances and ages: the closer they are, the older and deader they are, so the Milky Way has stunted their growth by stealing their gas. The nearest dwarf spheroidals, Ursa Minor and Draco, which are 220,000 and 250,000 light-years away, consist entirely of old stars, having stopped star formation long ago; but the farthest dwarf spheroidal, Leo I, 740,000 light-years away, illuminates itself with some stars born just a few billion years ago, indicating it retained gas until much more recently. The Milky Way's intrusion into its satellites' lives can hardly be unique. Other giant galaxies, such as Andromeda, must have extracted a price from their colonies, too.

Like the Milky Way, Andromeda rules an empire of satellite galaxies. This empire may even include the delicate spiral M33, the third largest member of the Local Group. M33 lies 2.6 million light-years from the Milky Way and 700,000 light-years from Andromeda. Although only a fifth of the Milky Way's luminosity, M33 abounds with molecular clouds that churn out new stars to keep the galaxy a youthful blue.

Outside the empires of the Milky Way and Andromeda lie the independent members of the Local Group. The two brightest, IC 10 in Cassiopeia and Barnard's Galaxy in Sagittarius, are irregular galaxies that resemble the Large and Small Magellanic Clouds. The other independent galaxies are small irregulars or dwarf spheroidals.

THE ANDROMEDAN EMPIRE

The closest giant galaxy to our own, Andromeda is the only galaxy in the Local Group that outshines the Milky Way. It is about 60 percent larger and brighter. The Andromeda Galaxy lies 2.4 million light-years from Earth and is the most distant object the average person can see with the unaided eye.

Despite Andromeda's proximity, its spiral nature eluded detection for decades after the discovery of more remote spirals, because Andromeda's disk tilts more edge-on to our line of sight, hiding its spiral. Only in 1887, after astronomers photographed the galaxy, did they trace its spiral structure. Andromeda's disk, which houses stars young and old, glistens blue and white, since its brightest stars are short-lived blue supergiants, while Andromeda's central bulge glows yellow-orange, because there its stars are old and the most luminous are orange and red giants, descendants of longer-lived stars like the Sun.

Andromeda has had a more glorious past than the Milky Way, because its bulge and central black hole are larger than their counterparts in the Milky Way.

FACING PAGE: A delicate spiral galaxy, M33 ranks number three in the Local Group, behind only Andromeda and the Milky Way.

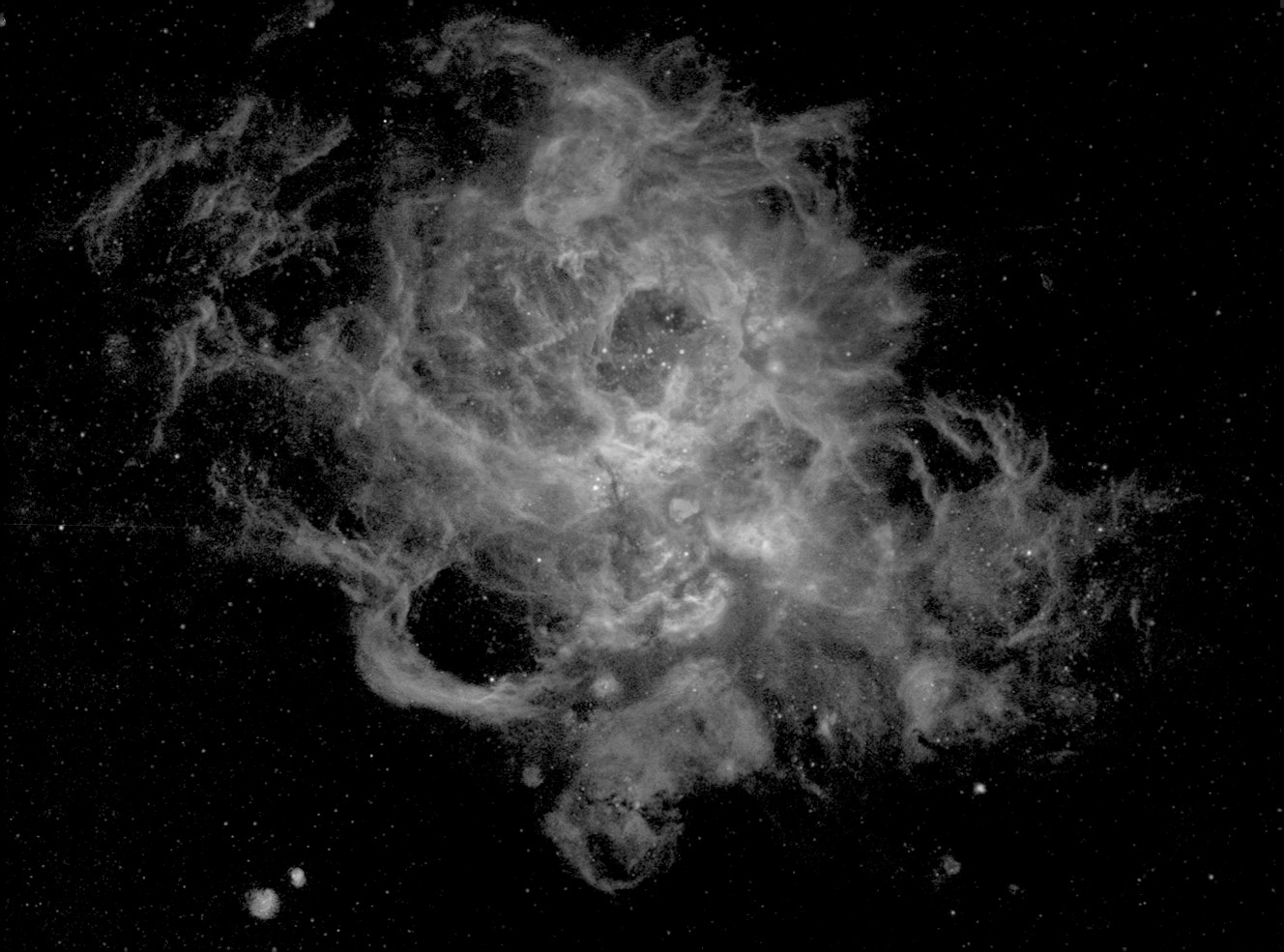

NEARBY GALAXY GROUPS

Just as the Milky Way is not the only galaxy, so the Local Group is not the only galaxy group. The nearest group to the Milky Way, 7 million light-years away, is the Sculptor group, named after an obscure constellation that houses most of the group's residents. The Sculptor group lies due south of the Galactic plane, so its inhabitants could see our Galaxy as a beautiful face-on spiral—a privilege we lack. The king of the Sculptor group, the edge-on spiral galaxy NGC 253, is smaller than the Milky Way, but its many clouds of gas and dust pump out so many new stars that astronomers call it a starburst galaxy. NGC 253's inhabitants may not even know the Milky Way exists, since they would have to view it through their own galaxy's disk of gas and dust. A fainter Sculptor group member, the spiral NGC 300, could pass for a twin of the Local Group's M33. Because its disk faces us, any residents could easily see the Milky Way.

In contrast to the Sculptor group, another nearby galaxy group lurks behind our Galaxy's gas and dust, so only in recent years have astronomers unveiled it. Called the Maffei group, it lies about 10 million light-years from the Milky Way. It takes its name from the astronomer who in 1968 discovered one of its largest members, Maffei 1 in the constellation Cassiopeia. Maffei 1 may be the nearest giant elliptical to the Milky Way. The Maffei group also includes the giant face-on spiral IC 342. Were it not for intervening gas and dust, this galaxy would be one of the brightest and most famous spirals in the sky. During the 1990s, astronomers spotted several additional Maffei group members, including the barred spiral Dwingeloo 1, named for the telescope that revealed it.

North of the Galactic plane lies the M81 group, which takes its name from a big spiral near the Big Dipper whose luminosity matches the Milky Way's. Without using binoculars or telescopes, sharp-eyed observers have actually glimpsed M81, making this galaxy, 12 million light-years distant, the farthest object visible to the unaided eye. In 1993, light from a supernova in M81 reached the Earth. Because of the galaxy's proximity, astronomers were able to examine preexplosion photographs and pinpoint the exact star that had blown up, only the second time this has been done; the first was for the 1987 supernova in the Large Magellanic Cloud. Near M81 is a smaller galaxy, M82, whose disturbed appearance once evoked images of an exploding galaxy. Today, however, astronomers classify M82 as a starburst galaxy, like NGC 253, that gives birth to a plethora of new stars. Its gas and dust probably got shook up when it passed its big neighbor.

FACING PAGE: M33 gives birth to new stars in nebulae like NGC 604.

FOLLOWING PAGES:

Page 140. NGC 253, an edge-on spiral and member of the Sculptor group, is the nearest starburst galaxy.

Page 141. Sculptor group galaxy NGC 300 resembles the Local Group's M33.

Page 142. Discovered only in the 1990s, the nearby galaxy Dwingeloo 1 would appear as a bright barred spiral—if it weren't hidden by the Milky Way's gas and dust.

Page 143. M81, a giant spiral, rules a galaxy group near the Big Dipper.

Page 144. M81's gravity has triggered a starburst in neighboring M82.

Page 145. Centaurus A is the nearest radio galaxy.

THE CENTAURUS group hosts the nearest radio galaxy, Centaurus A

galaxy's copious radio waves signal a disturbance. Centaurus A's main body appears to be that of a giant elliptical, but a thick lane of dust runs across it, possibly the shredded remains of another galaxy that strayed too close. The Centaurus group also hosts the exquisite M83, a beautiful face-on Sc spiral in the neighboring constellation of Hydra. Like the Milky Way, M83 has a bulge that is not perfectly round but somewhat barred.

At greater distances lie other intriguing galaxies. The magnificent type Sc grand-design spiral M51, the Whirlpool Galaxy, was the first galaxy in which astronomers saw a spiral, back in 1845. That feat was achieved by Ireland's Lord Rosse, who took aim at the galaxy with what was then the world's largest telescope. M51's small neighbor galaxy, which probably stirs up the main galaxy and intensifies its beauty, will pay a price for its good deed: the companion circles M51 once every half billion years and will be swallowed by it after just two or three more orbits. Another beautiful face-on spiral, M101, lies near M51: both galaxies appear around the Big Dipper, M51 in Canes Venatici, M101 in Ursa Major.

FACING PAGE: New stars arise around Centaurus A's center.

CLUSTERS OF GALAXIES

Although most galaxies make their homes in groups, clusters stand out more dramatically. The nearest galaxy cluster, some 50 million light-years from Earth, is Virgo, named for the constellation where most of its members gather. At the Virgo cluster's heart is the giant elliptical galaxy M87, which shoots a jet of material across thousands of light-years of space. M87 harbors a whopping 13,000 globular clusters, far outnumbering the 147 known in our Galaxy, and M87's center is in the grip of a giant black hole that puts our Galaxy's central black hole to shame: whereas the Milky Way's black hole weighs several million solar masses, M87's weighs several *billion*. Evidence for this black hole comes from the high speed at which material near M87's center revolves around the galaxy.

The Virgo cluster harbors thousands of other galaxies, too—ellipticals, spirals, and irregulars. It is the hub of the Local Supercluster, a vast chain of galaxies that bridges some 100 million light-years. The Local Group and its neighbors—the Sculptor, Maffei, Centaurus, and M81 groups—reside at one end of the Local Supercluster, and other groups reside at the other end, beyond the Virgo cluster.

GALAXY EVOLUTION

Like the stars they shelter, galaxies change. The different environments that giant ellipticals and spirals inhabit offer one clue to their origin and evolution, as do the ages of their stars. Because it consists entirely of old stars, the typical elliptical passed its prime long ago. Early in the universe's life, the galaxy burst forth brilliantly, forming all its stars quickly, but then ran out of interstellar gas and dust, and could give birth to no more. Now the galaxy is fading as its remaining stars die. In contrast, a spiral like the Milky Way starts life more slowly, forming only some of its stars early on, thereby conserving its gas and dust so that it can continue star formation for billions of years afterward, sprinkling itself with stars that keep it young and beautiful. Since giant elliptical galaxies dominate rich clusters like Coma, and since such clusters must have arisen from dense concentrations of matter in the early universe, perhaps dense regions of the early universe preferentially give birth to giant ellipticals, whereas sparser regions of the early universe, which formed galaxy groups, give birth to giant spirals.

However, this may not be the whole story, because the galactic content of rich clusters has actually changed over the universe's life. Using powerful telescopes, astronomers can probe the past and see how galaxies, groups, and clusters evolve. Although they cannot examine the Milky Way or M87 when these galaxies were young, they can see young galaxies that are destined to become galaxies like them. Astronomers do this simply by looking far away. Light from a galaxy 5 billion light-years distant takes 5 billion years to travel to Earth, so the galaxy appears not as it is

NGC 1365, brimming with gas, dust, and newborn stars, sweeps across vast expanses of space

have long observed it; but south of the Galactic plane, at a similar distance from the Milky Way, lies another galaxy cluster, Fornax. Smaller and more compact than the Virgo cluster, Fornax nevertheless bears a giant elliptical—NGC 1399—that like M87 abounds with globular clusters. But the stand-out galaxy here is the majestic barred spiral NGC 1365, one of the most graceful galaxies in the entire cosmos. It resembles a giant letter S: two vibrant spiral arms brimming with gas, dust, and newborn stars sweep across vast expanses of space from either end of a barred bulge built of aging stars. The galaxy lies 60 million light-years from Earth.

The Virgo and Fornax clusters impress in part because they are close to Earth. In the same general direction as Virgo but five to six times farther looms a galaxy cluster that could swallow both Virgo and Fornax, the Coma cluster, which has several times more galaxies than Virgo and Fornax combined and harbors two cD galaxies, NGC 4889 and NGC 4874. Whereas the brightest galaxies in the Local Group are spirals, the brightest in Coma are ellipticals, illustrating how large ellipticals prefer crowded environments and large spirals emptier ones.

FACING PAGE: Few clusters boast a gem like the Fornax cluster's NGC 1365, a majestic barred spiral galaxy.

today but as it was 5 billion years ago, before the Sun and Earth were born; and a galaxy 10 billion light-years from Earth appears younger still. The sky therefore resembles an archaeological dig: by digging deeper into the ground, an archaeologist reaches artifacts from longer ago, and by looking farther into the sky, an astronomer reaches galaxies from further into the past.

Distant galaxy clusters illuminate how at least some galaxies evolve. In a rich, nearby cluster like Coma, spirals like the Milky Way constitute a distinct minority, making up a mere 5 percent of the bright galaxies. Most of the rest are giant ellipticals. In contrast, rich clusters 5 billion light-years away, and thus 5 billion years younger than Coma, boast a spiral fraction of 30 percent. This indicates that rich clusters once had more spirals than they do today and so have become more elliptical-rich over time; therefore, some spirals may have become ellipticals. They did so by smashing into each other.

COLLIDING GALAXIES

Collisions play a much greater role in a galaxy's life than in a star's life. Few stars ever experience a collision, whereas few large galaxies escape one. In our part of the Milky Way, stars are so far apart that they never collide: shrink the Sun to the size of a quarter and on the same scale the nearest star would be four hundred miles away. But shrink the Milky Way's disk to the size of a quarter and its nearest large neighbor, Andromeda, would be just a foot and a half away, and the Milky Way's satellites mere inches away. No wonder, then, that galaxies frequently bump into one another.

Strangely, for decades astronomers failed to recognize this, believing instead that most galaxies live their lives in splendid isolation. Furthermore, no one foresaw the devastation that galactic collisions produce, because astronomers knew that stars are so widely separated that even during such a traumatic event, few if any stars from one galaxy collide with those from the other. Consequently, old paintings of colliding galaxies depicted two spiral galaxies passing right through each other, undamaged.

In 1972, however, Alar and Juri Toomre performed computer simulations of how one galaxy's gravity twists and distorts the other. The Toomre brothers matched their simulations to actual odd-looking galaxies, convincing other astronomers that these galactic

FEW STARS ever experience a collision, whereas few large galaxies escape one

peculiarities indeed arose from collisions. The Toomres' most celebrated example was the Antennae, two spiral galaxies some 80 million light-years from Earth that have run into each other in the constellation Corvus. A long winding tail of stars juts out of each galaxy, a consequence of galactic tides. Just as the Moon's gravity raises tides on both the near and far side of the Earth, because the Moon tugs more strongly on the near side and more weakly on the far side, so each galaxy's gravity raises tides on both the near and far side of its neighbor, yanking out a strand of stars. Such tidal tails are a hallmark of colliding galaxies.

Collisions can transform beautiful galaxies into ugly ones. Smash two delicate, ornate vases together and one is not left with a single ornate vase; one is left with a mess. Smash two delicate spiral galaxies together, each with stars on highly ordered orbits, and one does not end up with a new spiral galaxy. Stars get tossed about, the delicate spirals are torn asunder, and the two galaxies merge into an orbital jumble in which stars from both galaxies get shuffled together like a deck of cards. The stars' orbits get stretched into a chaotic mess of highly elongated paths, and there arises a giant elliptical galaxy, like M87. Many giant ellipticals, including M87 itself, may be the products of celestial collisions that occurred billions of years ago. Such collisions may have been especially common in newborn galaxy clusters, where galaxies are packed together. This may be one reason giant ellipticals proliferate in rich clusters.

Galaxy collisions affect stars and gas quite differently. Because a galaxy's stars are so small and spread out from one another, few if any collide, though all feel the intruder's gravity. The gas, though, faces calamity, because gas clouds span tens or hundreds of light-years. Clouds in one galaxy therefore smack those in the other. These collisions squeeze the gas, unleashing a flood of new stars. Collision-induced star formation can explain why giant ellipticals formed all their stars in a single burst.

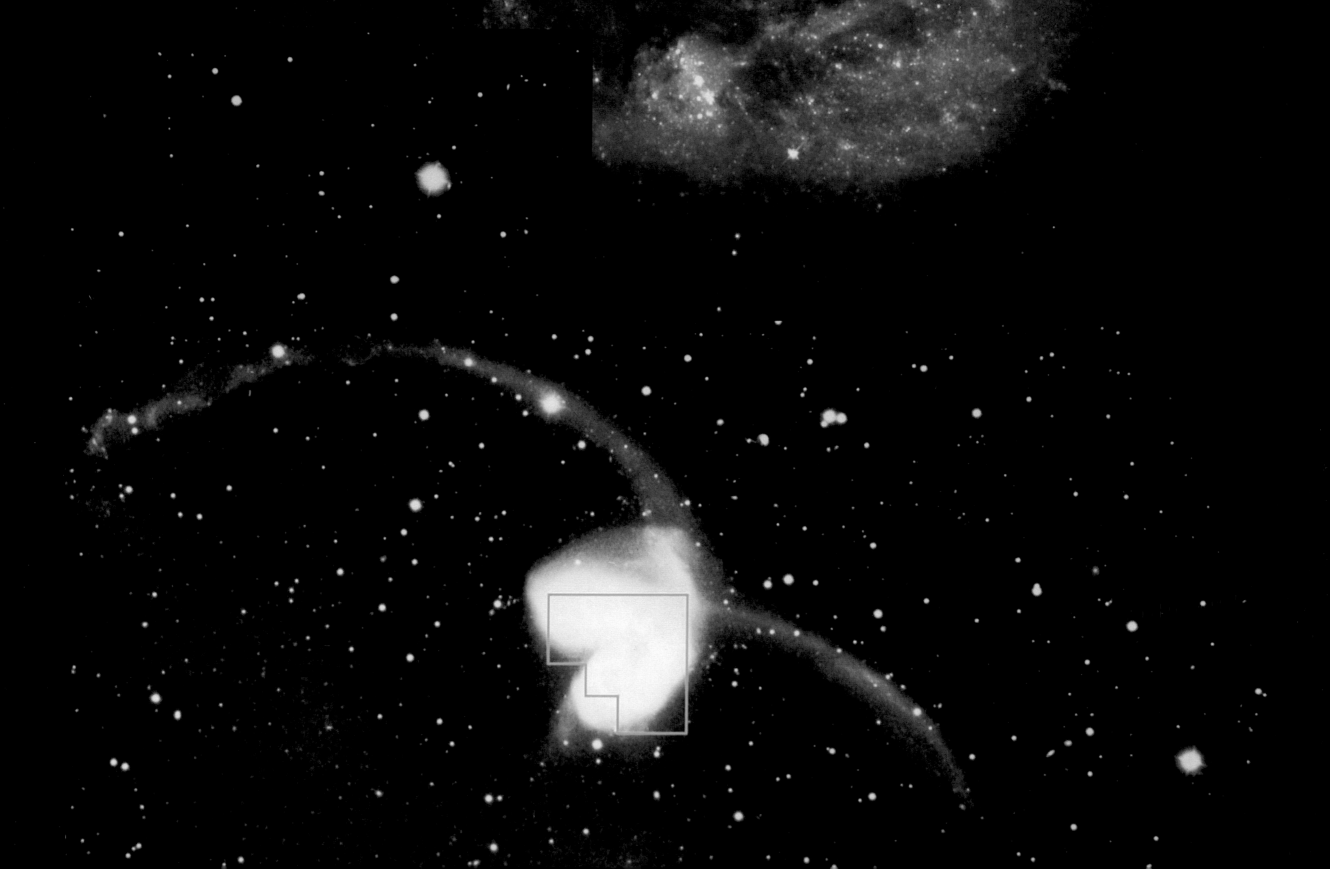

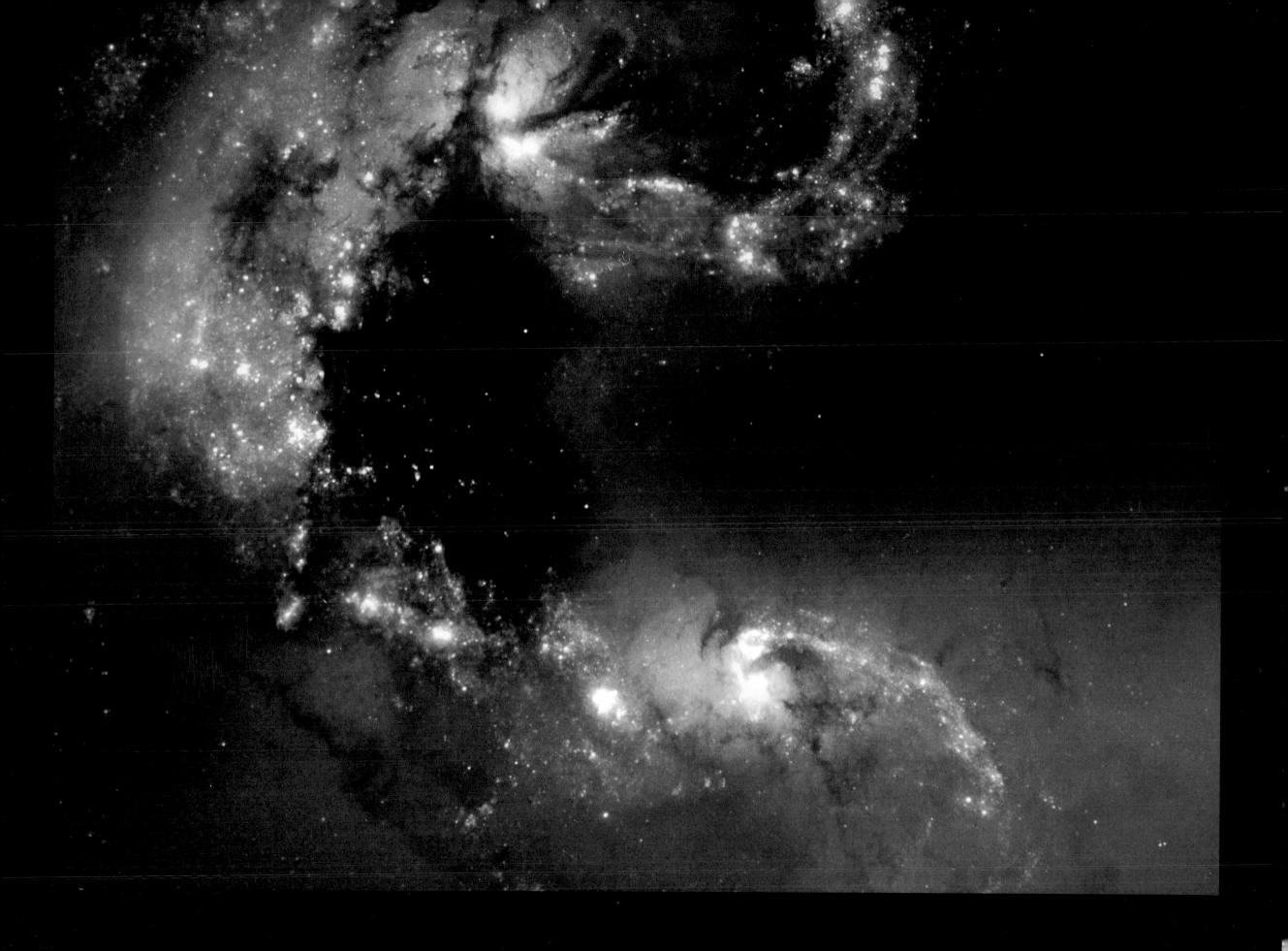

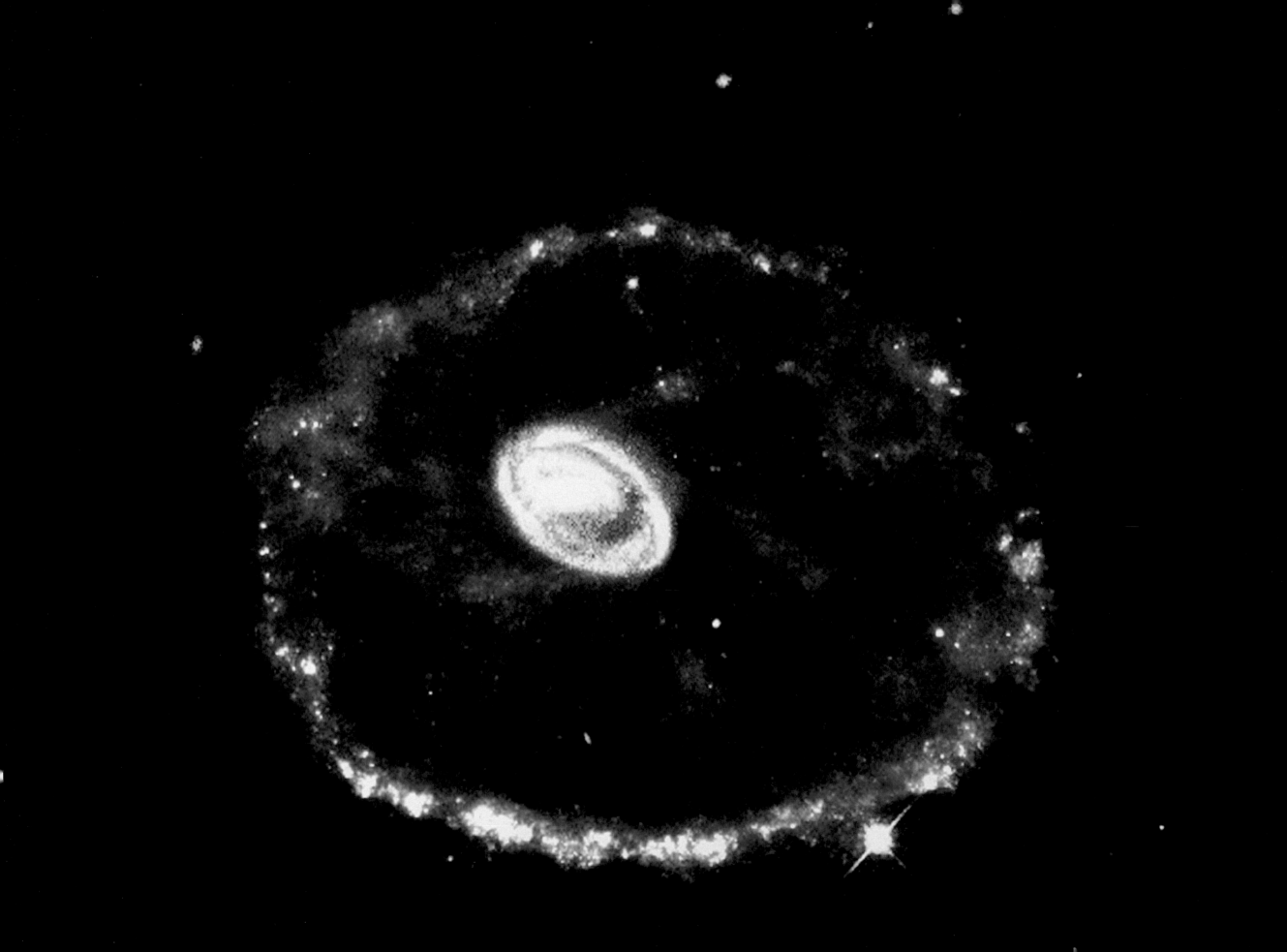

rals into ellipticals can actually enhance rather than detract from the universe's beauty. A special type of collision produces the exotic ring galaxy, which looks like a circle of starlight suspended in space. A ring galaxy forms when a small galaxy scores a bull's-eye, smashing into a spiral galaxy's center. Like a pebble plunging into a pond, the intruder's gravity generates a wave that races outward and compresses the spiral's gas, spawning new stars in a ring that expands away from the galaxy's center. The best-known ring galaxy, the Cartwheel, lies half a billion light-years from Earth and sports a ring slightly larger than the Milky Way's disk. Ring galaxies are rare, however, because only rarely does a small galaxy hit a large one head-on, and the ring itself soon fades as the wave of star formation dies down.

BILLIONS OF YEARS from now, the Milky Way may collide with Andromeda

Galaxies collide and interact even within the Local Group. In 1994, astronomers in England discovered the ruins of a dwarf spheroidal galaxy in the constellation Sagittarius some 80,000 light-years from the Sun. The Milky Way has stretched this galaxy out like taffy, and the stars and star clusters of the Sagittarius dwarf will merge with the Galaxy's stellar halo. Indeed, Sagittarius carries what astronomers had previously recognized as the second most luminous globular cluster in the Milky Way, M54. Likewise, the Milky Way may have acquired many of the stars and globulars now lighting its stellar halo by looting other small galaxies.

And billions of years from now, the much more substantial Magellanic Clouds will spiral into the Milky Way. Unlike the Sagittarius dwarf, the Magellanic Clouds bear large amounts of interstellar gas, so the collision will hurl Magellanic gas clouds against the Milky Way's, compressing the gas, triggering an avalanche of new stars, and reinvigorating the Milky Way by turning it into a starburst galaxy, when it may outshine its great rival, Andromeda.

Even as it augments its own power, the Milky Way may face a threat itself. Billions of years from now, it may collide with Andromeda, and the two giants will merge into a giant elliptical galaxy that would be the largest and brightest for millions of light-years around. At the moment, though, no one can say whether such a collision will materialize. Astronomers have long known that every day Andromeda creeps 6 million miles closer to the Milky Way, but Andromeda's sideways velocity is unknown. If that is large, then Andromeda will miss the Milky Way and simply dance around it. On the other hand, if Andromeda's sideways motion is small, then the galaxy is heading straight toward us, and a collision is inevitable. Only when astronomers measure Andromeda's sideways velocity, which they may do in the next two decades, will they know our Galaxy's ultimate fate.

Galaxy collisions occur both near and far. In the distant reaches of the universe, collisions may spark the ultraluminous entities that flourished when the cosmos was young: quasars.

giant black holes, but none is a quasar. Instead, a colli-sion between two gas-rich galaxies may be required to stoke the black hole. When the galaxies collide, their gas clouds hit one another, lose orbital energy, sink toward the galaxy's center, and whirl around the black hole, igniting the quasar.

BLACK HOLES power the greatest displays of light in the universe

Quasars are galaxies at the extreme, pouring out more light than any other type. When astronomers first found quasars, in 1960, they thought them mere stars in our Galaxy, peculiar only because they emitted radio waves. ("Quasar" means "quasi-stellar radio source.") In 1963, however, Maarten Schmidt at the California Institute of Technology confronted these monsters face to face: he discovered that a quasar named 3C 273 lay *billions* of light-years from Earth. In order to be seen from such a great distance, the quasar must be extra-ordinarily luminous, radiating trillions of times more light than the Sun and hundreds more than the Milky Way. And 3C 273 was not alone. Other quasars were remote, too, and today astronomers have found some that shine from a distance of 10 billion light-years—almost to the edge of the observable universe. Astonishingly, whereas the Milky Way's light comes from a disk 120,000 light-years across, a quasar's much more intense radiation floods out of a region smaller than the solar system. Astronomers know this because they see some quasars flicker wildly, in just a day's time, indicating that most of the light originates from a region less than one light-day across, about twice the diameter of Pluto's orbit.

Ironically, such a luminary owes its brilliance to what may seem its opposite: a giant black hole, millions or billions of times more massive than the Sun, that sits at the center of a galaxy. The black hole grabs gas,

Evidence that quasars arise when galaxies collide emerged during the 1980s, after NASA launched the Infrared Astronomical Satellite (IRAS). Infrared radiation carries less energy than visible light. Stars emit it, but so do objects too cool to shine, such as planets and people. Interstellar dust also radiates infrared, because dust gets heated by starlight. IRAS revealed that some odd-looking galaxies, which owe their peculiarities to collisions, release as much energy at infrared wavelengths as quasars do at visible ones. These objects may be quasars so shrouded in dust that the quasar's light gets trapped and heats the dust, heat that escapes at infrared wavelengths.

The collision that stoked the monster black hole and lit the quasar also breeds hordes of new stars. As these stars arise, their winds and explosions poke holes through the dust, so eventually visible light from the quasar streams out and reaches the Earth. Right now, for example, the ultraluminous object Markarian 231 is transitioning from infrared to optical quasar. In 1986, astronomers discovered it is really a colliding galaxy, for two tidal tails, similar to the Antennae's, protrude out of it. Markarian 231 will continue to evolve. As the dust further dissipates, the infrared luminosity will decrease and Markarian 231 will become a full-blown optical quasar like 3C 273.

BECAUSE MOST quasars are extremely distant, they existed long ago

The quasar will not shine forever, though. Instead, the gas fueling the quasar will get depleted as the black hole swallows it and as new stars arise from it. Then the quasar will die. In its place will be a dormant black hole throned at the center of a large elliptical galaxy. When we look at M87, therefore, we may be seeing a burned-out quasar, which once outshone nearly all else in the cosmos.

If quasars occur when two large galaxies collide, and if each galaxy has a large black hole, then some quasars should possess binary black holes—two giant black holes orbiting each other. The quasarlike object OJ 287, in the constellation Cancer, appears to have such a black hole binary. Its black holes revolve around each other once a decade on an elliptical orbit. Consequently, every decade the two black holes approach each other, one black hole's gravity disturbs the other black hole's disk of orbiting gas, and OJ 287 flares up.

Because most quasars are extremely distant, they existed long ago, when the universe was young. The universe is expanding, so long ago it was smaller and more crowded, and galaxies collided more often. In addition, when the universe was young, galaxies had only just started to transform gas into stars, so galaxies had more gas to feed black holes and power quasars. Thus, quasars thrived long ago, when galactic collisions were more frequent and galaxies more gas-rich.

That quasars once shone prolifically and now have largely vanished demonstrates one way in which the cosmos has changed since its youth. Astronomers who seek to understand the overall universe, however, venture back beyond even the quasar epoch, to a time before stars and galaxies and quasars shone—to the fiery birth of the cosmos, the big bang. To the universe, then, we now turn.

THE UNIVERSE

The planets, stars, and galaxies voyage like flotsam on the texture of the universe. Cosmologists, who study the universe's grand design, ask questions that once resided in the realms of philosophy and theology: Is the universe finite or infinite? If finite, where is its center and where is its edge? Did the universe have a beginning, or has it existed forever? If it had a beginning, how was it born, and what happened before? What will be its fate—will it live forever or someday die? Are there other universes besides ours? And why is there a universe at all? Does it have some purpose, or is it just an accident?

Cosmology abounds with controversy. Observations that support various theories are often scant or nonexistent. Furthermore, there is only one known universe, so all cosmological experiments try to characterize a single entity, whereas astronomers studying planets, stars, or galaxies have numerous objects on which to test ideas. Moreover, cosmological questions—such as whether the universe had a beginning—carry philosophical and religious implications that scientists may like or dislike regardless of the actual data.

AND WHY is there a universe at all?

FACING PAGE: Cosmology's first observation: the night is dark.

astronomer but by American poet and
Allan Poe. Suppose, said Poe, that the univ
existed forever. Then even if the universe i
night will still be dark, because light from
galaxies has not yet had time to reach th
example, if the universe is 12 billion years
Earth can receive light only from objects
lion light-years. Thus, there is not an infini
spherical shells contributing light to the
sense, then, the dark night sky testifies
verse's creation.

It also testifies to the universe's
1964, cosmologist Edward Harrison, armed
ern knowledge of how stars shine and e
lated that the universe would need 10 t
more stars than it has in order to look brig

'S A SMALL UNIVERSE AFTER ALL

Despite its grandeur, the cosmos is actually smaller than most people think. The observable universe stretches 10 to 15 billion light-years in all directions from Earth, but that sounds more impressive than it is. Shrink the Sun to the size of the period at the end of this sentence, and on the same scale our Galaxy would be far larger than the entire world; but shrink our *Galaxy* to the size of that period, and the observable universe would be only ten times larger than your living room.

E DARK NIGHT SKY

The first cosmological observation is both simple and profound: the sky is dark at night. Yet if the universe is infinite, filled with stars and galaxies, the amount of light reaching Earth should make the night blazingly bright.

This contradiction between theory and observation is called Olbers' paradox, after German physician and astronomer Heinrich Wilhelm Olbers. He discovered two of the first four asteroids and in 1823 published a paper on the dark night sky that he is best known for today, even though others had raised the issue before. Olbers' paradox works like this. Imagine an infinite universe uniformly filled with stars, or stars clustered into galaxies. Galaxies nearby look brightest, but distant ones also contribute light. These distant galaxies are more numerous, simply because there is more space at greater distances, and it turns out that their greater number exactly compensates for their fainter appearance, so each spherical shell of space around Earth contributes an equal amount of light to the sky. Since by assumption the universe is infinite, an infinite number of these spherical shells exist, each sending an equal amount of light to Earth, so the sum should be infinite. Yet the night is dark.

Although the night sky appears serene, the universe is not static but expanding. This startling discovery ranks as the greatest cosmological find of the twentieth century. The first evidence of cosmic expansion emerged in the 1910s, from the same man and indeed the same data that suggested the "spiral nebulae" were galaxies outside the Milky Way: Lowell Observatory's Vesto Slipher, who was examining these objects because of their supposed connection to solar systems. In 1912, through painstaking effort, Slipher recorded the spectrum of what is now called the Andromeda Galaxy. A spectrum reveals a galaxy's speed, through the Doppler shift. When a speeding siren approaches, its sound waves get scrunched to shorter wavelengths, so its pitch sounds higher; when the siren races away, its sound waves get stretched to longer wavelengths, so the pitch is lower. The same thing happens to light. Light waves from an approaching galaxy get scrunched to shorter wavelengths, a phenomenon called a blueshift, since blue light has a short wavelength, and light waves from a receding galaxy get stretched to longer wavelengths, which is called a redshift.

Slipher discovered the Andromeda Galaxy to have a large blueshift, indicating that the galaxy moves toward us fast. Slipher then observed other spirals. These had large velocities, too, but unlike Andromeda most had redshifts, which meant they were moving away from Earth. Furthermore, the fainter and presumably farther galaxies tended to have greater redshifts and so were moving away faster, the key observation that astronomers now interpret as the sign of an expanding universe.

However, Slipher never measured these galaxies' actual distances. That task fell to Edwin Hubble at Mount Wilson Observatory in California, and in 1929 Hubble announced his famous result: the farther a galaxy, the greater its redshift. Although in his first paper Hubble did not attribute the distance-redshift relation to an expanding universe, that is exactly how cosmologists view it today. In fact, contrary to popular belief, the redshifts do not even arise from the Doppler shift. Galaxies do not move through space away from Earth; instead, the space between them and Earth

expands and carries them away. Thus, in the modern conception, space is something rather than nothing. As a galaxy's light travels through this expanding space, the light's wavelength gets stretched. The longer the light has traveled—that is, the farther the galaxy from Earth—the more the light has been stretched; so the farther the galaxy, the greater is its redshift, which is just what Hubble found. The galaxy indeed moves away from us, because the expanding universe forces it to, but this movement does not produce the redshift. Rather, both the redshift and the galaxy's recession arise from the expansion of space. The incorrect concept is:

Galaxy moves away from us ——> galaxy exhibits redshift.

The correct sequence is:

Space between us and galaxy expands

galaxy exhibits redshift.

galaxy moves away from us.

Astronomers distinguish this redshift from a Doppler shift by calling it the cosmological redshift.

To visualize the universe's expansion, imagine a balloon onto which a few coins have been firmly glued. The balloon's skin represents space; each coin is a galaxy. Call one coin the Milky Way. Now blow up the balloon and its skin expands, carrying every other coin away from the Milky Way. A coin close to the Milky Way moves little, but a distant coin moves more, just as distant galaxies have larger redshifts than nearby ones. Even though all coins move away from the coin representing the Milky Way, it is not at the center of the balloon, just as the real Milky Way is not at the center of the universe.

In Hubble's 1929 paper, the wavelengths from the farthest galaxy he mentioned were only 1 percent longer than they were when emitted, so their redshifts were just 0.01. The Coma cluster's redshift is twice this, 0.02, which means its light gets stretched 2 percent by the expansion of space. The quasar 3C 273, the first to have its approximate distance measured, is at redshift 0.16, which corresponds to a distance of some 2 billion light-years. Astronomers first exceeded redshift 1 in the 1960s. Redshift 1 means a 100 percent increase in wavelength—the wavelength has doubled. Thus, when an object with redshift 1 emitted its light, the universe was only half as big as today; since then, the universe's size has doubled, doubling the wavelength. Astronomers also broke the redshift 2 barrier in the 1960s, redshift 3 in the 1970s, redshift 4 in the 1980s, and redshift 5 in 1997. An object with redshift 5 sent its light toward Earth when the universe was just one-sixth of its present size.

In this analogy, the balloon's two-dimensional skin corresponds to the universe's three-dimensional space, and everything astronomers see lies *within* this skin. Therefore, the following questions, which popular books sometimes pretend to answer, have no known answers: What is the universe expanding into? (Popular books sometimes say "nothing," but since astronomers can't observe outside the universe, no one knows.) Does the universe have a center? (Popular books say no, and indeed no point within the universe is its center, just as no point on the balloon's skin is the balloon's center; but since astronomers can't see outside the universe, no one can say whether there is some center outside the universe.) Are there other universes? (Since astronomers can't see them, no one knows.)

Because the universe expands, people sometimes incorrectly think that everything is moving away from Earth—that all stars are receding, and all galaxies, too. In fact, a galaxy's gravity more than overpowers space's expansion, so the Milky Way gets no bigger and its stars no more widely separated, just as the coins did not get bigger when the balloon expanded. Likewise, several nearby galaxies, such as Andromeda, move toward the Milky Way, because the gravitational attraction between them and the Milky Way overcomes the expansion of space.

Astronomers quantify a galaxy's redshift by comparing a wavelength of light that reaches Earth with the wavelength it had when it left the galaxy. The greater the difference between the two wavelengths, the greater is the redshift. For example, a galaxy normally has two prominent spectral lines from calcium that appear purple, because calcium absorbs purple light and imprints dark lines onto the purple portion of the galaxy's spectrum. A redshift can cause these spectral lines to look blue; a greater redshift makes them look green and then yellow; and still greater redshifts take them to the orange, the red, and finally the infrared, wavelengths longer than those of visible light.

THE OBSERVABLE UNIVERSE

The universe we see is not the one we inhabit. That is because light does not travel infinitely fast, so as we look up, we see the universe as it was, not as it is. Scientists and philosophers once debated whether the speed of light was infinite. French mathematician and philosopher René Descartes argued that it must be infinite, for otherwise light from different objects would reach us at different times, complicating observations of the universe. But complicated they are.

For nearby objects, this delay is no big deal. Moonlight takes just a second to reach Earth, so we see the Moon as it was a second ago, during which time the Moon has hardly changed. Likewise, we see the Sun as

ARE THERE other universes?

During the 2.4 million years that light from the Andromeda Galaxy takes to reach us, thousands of its stars have exploded as supernovae, and millions of other stars have been born, but these are small numbers compared with the several hundred billion stars that illuminate Andromeda. At much greater distances, though, galaxies can change substantially during their light's lengthy journey to Earth. Quasars, which thrived shortly after the universe's birth, have largely disappeared. Were we to travel to one, we would find that by the time we got there, it would have faded and turned into a giant elliptical galaxy like M87. Right now, distant astronomers studying quasars may be looking at M87 itself, which to them but not us appears as a quasar.

The finite speed of light both hinders and helps astronomers. On the negative side, they can never see the universe as it is, and they can never, even with their best instruments, see the whole universe. They can see no farther than the universe is old. If the universe is, for example, 12 billion years old, they can see no farther than 12 billion light-years, even if the actual universe is much bigger. And the farther astronomers look, the more what they see differs from the present reality. On the positive side, the finite speed of light preserves the past. Were the speed of light infinite, we would see few if any quasars, because few if any exist today; we know of their existence only because space has preserved the light beams they sent out before their deaths. In the same way, we know of the existence of dinosaurs because the land has preserved some of their remains as fossils. In principle, astronomers can even see back to creation itself, by peering far enough.

expands. When astronomers say an object is, for example, 8 billion light-years from Earth, they mean the object's light took 8 billion years to reach Earth; but because the universe is expanding, that object must have been closer when it emitted the light, and must be farther now, because the object has moved away during the time the light traveled here.

THE BIG BANG

Because the universe is expanding, it was once smaller, and 10 to 15 billion years ago the whole cosmos was compressed into a point. This point burst forth in a big bang and gave birth to the universe. The big bang was *not* an explosion of matter into preexisting space. Instead, it marked the birth of space itself, which at that time began to expand.

The big bang theory means the universe had a beginning, which some people find appealing and others appalling. Atheists, in particular, dislike a beginning, for it sounds like the work of a supernatural being. In 1948, astronomers conceived an alternative cosmology, the steady state theory, which holds that although the universe is expanding, it has existed forever. This theory says that as galaxies drift away from one another, new galaxies are born, because fresh particles pop out of space. Although this "continual creation" struck some scientists as absurd, the postulated rate of creation was so low that it violated no experiment, and steady state supporters argued that single atoms popping into existence were much less absurd than an entire universe doing so. Unlike the big bang theory, however, the steady state theory had to confront cosmology's oldest observation—the darkness of the night sky, Olbers' paradox. The big bang theory can say the night is dark because the universe had a beginning, and light from beyond a certain distance hasn't had time to reach Earth. But if the universe had no beginning, as the steady state theory maintained, then the universe was infinitely old, so why isn't the night bright? Salvation for the steady state lay solely in the expansion of space, which causes extremely distant galaxies to recede from Earth faster than light, so their light never reaches us and the night stays dark.

But the steady state theory proved wrong. During the 1950s, astronomers began to realize that the universe contains far more helium than stars can make, indicating that the element arose elsewhere—in the big bang. Astronomers also found that radio galaxies proliferate at great distances, so more existed long ago than today, which contradicts the steady state view that the universe remains basically the same. Likewise, the 1960s discovery that most quasars existed long ago contradicts steady state cosmology, too. But the death blow to the steady state came in 1965, when radio astronomers stumbled across whispers of the universe's birth.

AFTERGLOW

If the universe began in a big bang, the heat of that explosion should still pervade the universe, and in 1948 big bang proponents Ralph Alpher and Robert Herman, working under George Gamow, calculated the temperature this afterglow should have. It was low, just a few degrees above absolute zero, because the universe's expansion has stretched the radiation's wavelength enormously and greatly reduced its energy. But the afterglow was there, a prediction that distinguished the big bang model from the steady state. Yet other astronomers forgot about it.

In 1965, Arno Penzias and Robert Wilson of Bell Labs in New Jersey detected this radiation. For some time they had been struggling to use a radio antenna to pick up radio waves from space, but a persistent hiss interfered. When other scientists heard of this problem, they recognized it as the background radiation from the big bang. It is called the cosmic microwave background, because much of its energy appears as microwaves, which are longer than infrared waves but shorter than most radio waves.

Since then, astronomers have measured the cosmic microwave background precisely. Its temperature is just 2.7 Kelvin, or -455 degrees Fahrenheit. It represents radiation from a redshift of 1,100, when the universe was only a few hundred thousand years old. At that time, the universe's light finally escaped the clutches of matter. To see why, journey back to the very early universe. In the first few seconds after the big bang, the universe was billions of degrees hot. As the universe expanded, however, it cooled, like an expanding gas. After a few minutes of expansion, the universe had cooled to millions of degrees; and then, after a few years, to many thousands of degrees. The intense heat prevented electrons from joining protons. Instead, the electrons roamed about freely, and free electrons interact strongly with light, scattering it every which way. Thus, as long as the universe was hot, it was also opaque, and its light was trapped with the matter. As the universe continued to expand, however, it continued to cool. When the universe grew to be a few hundred thousand years old, its temperature dropped to 3,000 degrees Kelvin (5,000 degrees Fahrenheit), and the electrons then combined with the protons to form hydrogen atoms—an event cosmologists call "recombination," a misleading term since it implies the elec-

IN THE FIRST few seconds after the big bang, the universe was billions of degrees hot

trons and protons had once been combined. Now, with the electrons trapped by protons, the light could stream through space unimpeded. It is this light, greatly redshifted by the universe's expansion, which Penzias and Wilson detected as microwaves.

In 1992, NASA's Cosmic Background Explorer (COBE) satellite discerned minute fluctuations across the sky in the cosmic microwave background. These resulted from fluctuations in the density of matter when the universe was a few hundred thousand years old and presumably sowed the seeds of cosmic architecture: the gravitational pull of dense regions attracted more matter and grew still denser, building the galaxies, clusters, and superclusters that crisscross the cosmos today, while emptier regions grew into the voids that bear few galaxies at all.

From time to time, NASA people who should know better put out grand claims that some new telescope will see farther than anyone has ever seen before. These statements are false. The farthest astronomers have ever seen is the cosmic microwave background, which comes from a time just a few hundred thousand years after the big bang. Prior to that epoch, the universe was opaque, and no telescope that detects any form of light will ever see farther. Nevertheless, cosmologists can probe back still further in time—by examining the elements created mere *minutes* after the big bang.

IGHT ELEMENTS

During the 1950s, astronomers realized that nearly all of the universe's elements—carbon, oxygen, iron, copper, silver, gold, and many more—arose in stars. Astronomers saw that old stars had fewer of these elements than young ones, indicating that the Galaxy had grown more enriched in these elements. A short-lived element, technetium, was found on red giant stars, which meant the stars themselves manufactured the element. And in 1957, Margaret and Geoffrey Burbidge, William Fowler, and Fred Hoyle published a mammoth paper detailing how the stars had created nearly every element on Earth. The paper starts with words from William Shakespeare's *King Lear*: "It is the stars, / The stars above us, govern our conditions."

But stars do not explain everything. Even during the 1950s, astronomers began to realize that one major problem was helium, the second lightest and second most abundant element, after hydrogen. Helium may seem an odd problem, since the Sun and most other stars create it. But the cosmos has five times more helium than stars could have produced. The rest arose in the fiery aftermath of the big bang, primordial nucleosynthesis. This era, which lasted just a few minutes, also gave birth to deuterium, a rare isotope of hydrogen that weighs twice as much as ordinary hydrogen, and to some lithium, the third lightest element.

The era of primordial nucleosynthesis did little else, however. Compared with stars, which assume various pressures and temperatures and therefore cook up a rich diet of elements, the early universe was simple, as was its menu. Because of this simplicity, nuclear physicists can predict exactly how much of each element and isotope formed. The predicted abundances depend on how dense the universe was, since the greater the density, the faster the nuclear reactions occurred, enhancing some elements and destroying others. These predictions can then be compared with the actual big bang abundances, preserved in old stars that formed shortly after the big bang and in small galaxies that possess few stars to alter the original composition. Remarkably, a single number for the universe's density yields the observed abundances of four different nuclei relative to hydrogen-1: deuterium, helium-3, helium-4, and lithium-7. This concordance provides additional evidence for the big bang.

Three Cosmological Parameters

For all the complexity of the planets, stars, and galaxies within it, the universe itself is sufficiently simple that its overall properties depend on just three parameters: the Hubble constant, which expresses how fast the universe expands and indicates its approximate age; omega (Ω), which measures the universe's density of matter, slows its expansion, dictates its fate, and like the Hubble constant affects its age; and lambda (λ), the cosmological constant, which if it exists accelerates the universe's expansion, influences its fate, and like the Hubble constant and omega affects its age. Still, all three parameters—the Hubble constant, omega, and lambda—are notoriously difficult to measure.

The Hubble Constant

No number in cosmology has been fought over more than the Hubble constant. Its basic utility, however, is simple: it converts redshift into distance. Without it, astronomers know only relative distances, not absolute ones. For example, whatever the numerical value of the Hubble constant, an object at redshift 0.02 is farther than one at redshift 0.01, but the Hubble constant states just how far each object is. The Hubble constant is especially critical because it affects the universe's age. The higher the Hubble constant, the faster the universe is expanding and the younger the universe must be, because a fast-expanding universe has taken less time to go from the zero size it had at the big bang to its present size.

For all its importance, the Hubble constant is just the redshift divided by the distance. It is expressed in the odd units of kilometers per second (the redshift, given as a Doppler shift) per megaparsec (a unit of distance equal to 3.26 million light-years, a bit farther than the Andromeda Galaxy). For example, if the Hubble constant is 65, then galaxy redshifts typically increase by 65 kilometers per second for every megaparsec farther one looks.

In principle, obtaining the Hubble constant is easy: just measure a galaxy's redshift, divide by the galaxy's distance, and *voilà*, you've got the Hubble constant. With modern equipment, the first task—measuring the galaxy's redshift—is indeed easy, but not so the second, measuring the distance. Furthermore, a galaxy gets tugged by the gravity of other galaxies near it, a tug that produces a Doppler shift which adds to or subtracts from the galaxy's cosmological redshift. For example, because it feels the Milky Way's gravity, the Andromeda Galaxy, which should be moving *away* from our Galaxy at about 50 kilometers per second, actually moves *toward* it at 120 kilometers per second. Consequently, in order to probe the universe's expansion and measure the Hubble constant, astronomers must deduce the distances of *distant* galaxies, whose cosmological redshifts are so large they overwhelm random Doppler shifts.

Measuring distances, even of nearby celestial objects, has never been easy. After all, astronomers can hardly travel to them with a yardstick. Only in 1838 did astronomers begin to measure the distances of nearby stars, by detecting tiny shifts in their apparent positions. These shifts, called parallaxes, result because the Earth moves around the Sun, so astronomers view the stars from different perspectives during the course of a year. The nearer the star, the larger is its parallax. Unfortunately, the parallaxes of stars more than a few hundred light-years distant are so small they can't be measured. Instead, astronomers must

NO NUMBER has been fought over more than the Hubble constant

FACING PAGE: In 1994, the war over the Hubble constant reached the Virgo cluster's M100, where astronomers detected Cepheids and established a

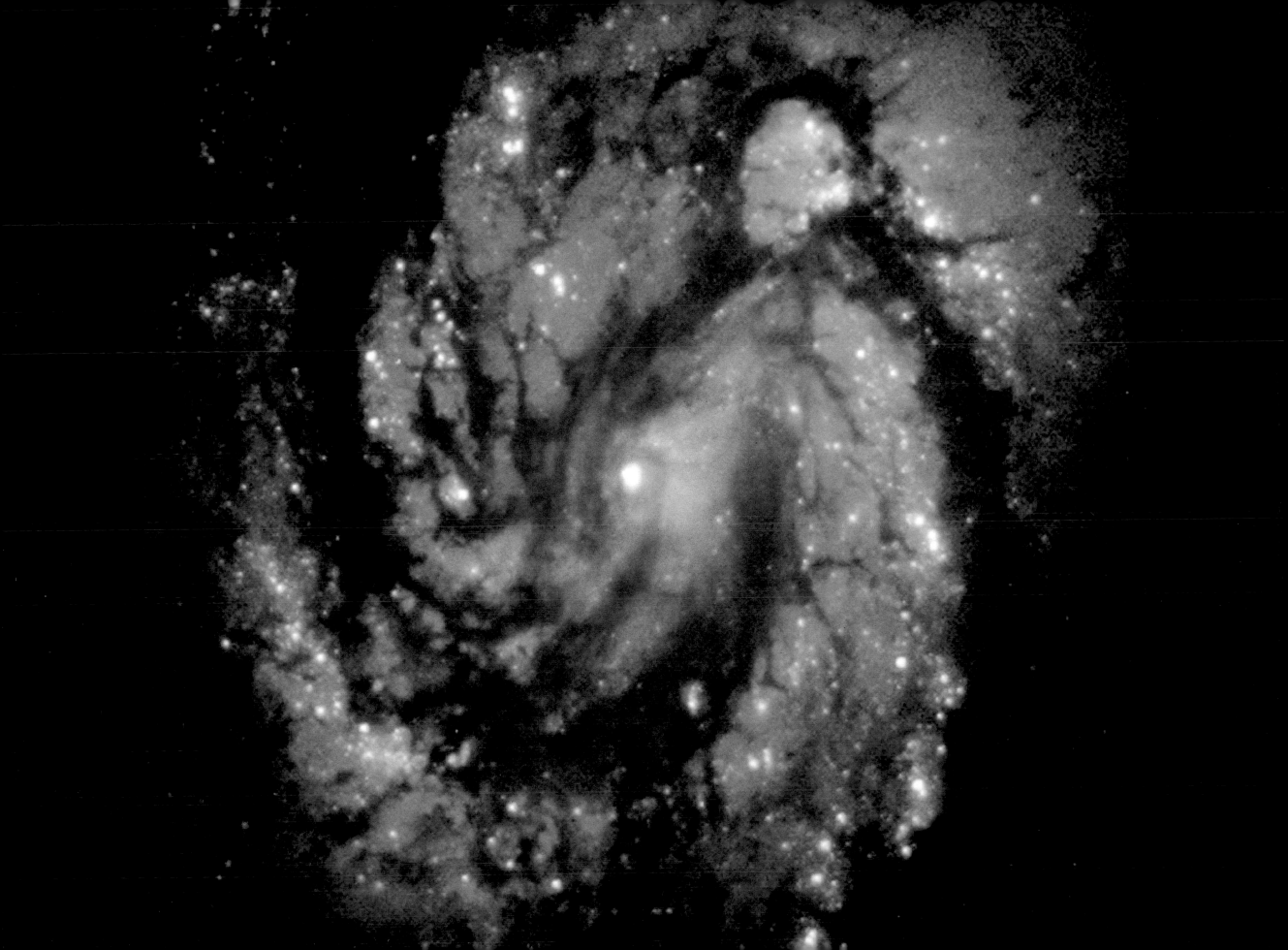

finding values for the Hubble constant between those of archrivals Sandage and de Vaucouleurs.

Before the launch of the Hubble Space Telescope, astronomers could see Cepheids only in nearby galaxies—Local Group members plus a few in close-by groups like Sculptor and M81. But the Hubble Space Telescope has spotted Cepheids in much more distant galaxies, such as the spiral galaxy M100 in the Virgo cluster and the barred spiral NGC 1365 in the Fornax cluster. M100's Cepheid distance is 53 million light-years; NGC 1365's Cepheid distance is 60 million light-years. As a result of these accomplishments, measured values of the Hubble constant have converged between 50 and 80. A middling value—65—implies a universe that began expanding 10 to 14 billion years ago. Furthermore, improved models of how stars evolve indicate that the globular clusters are several billion years

THE UNIVERSE, as it should be, is older than its oldest stars

younger than had been thought, with ages of 12 to 14 billion years. Thus, within the uncertainties, the age conflict between the stars and the universe no longer exists. The universe, as it should be, is older than its oldest stars.

However, the universe's age depends not just on the Hubble constant but also on omega, a number even more uncertain and even more important: it determines the fate of the cosmos.

infer the distances in other ways—for example, by assuming that a distant star of a particular type has the same intrinsic brightness as a nearby star of the same type. Comparing the distant star's apparent brightness with its assumed intrinsic brightness then yields the star's distance.

To fathom the distances of entire galaxies, astronomers deploy an arsenal of distance indicators. The most important are the Cepheids, the same pulsating yellow supergiant stars that Edwin Hubble employed to establish the extragalactic nature of the Andromeda Galaxy. The longer a Cepheid takes to pulsate, the more luminous it is; comparing this luminosity with the Cepheid's apparent brightness reveals the distance of its galaxy.

Hubble was the first to try to measure the constant that bears his name. He relied on Cepheids to give him distances to the nearest galaxies, such as Andromeda and M33, but beyond those galaxies he couldn't see Cepheids, so he turned to less reliable distance indicators. When he finished, he wound up with a Hubble constant ten times too high. It meant the universe expanded so fast it was younger than Earth—clearly impossible.

After Hubble's death, in 1953, Allan Sandage carried on the burden. Eventually Sandage teamed up with Swiss astronomer Gustav Tammann and in 1975 concluded that the Hubble constant was around 55. This implied a slow expansion for the universe and thus an old age, between 12 and 16 billion years.

In 1976, however, the modern Hubble war erupted. French-born Texas astronomer Gérard de Vaucouleurs said Sandage and Tammann were wrong and that the Hubble constant was nearly twice what they had said, around 100, implying a universe only half as old, 6 to 8 billion years. Such a young age posed a major problem: it meant the universe was younger than the Milky Way's globular clusters, then thought to be around 15 billion years old. After de Vaucouleurs' challenge, other astronomers entered the fray, many

OMEGA: THE WEIGHT—AND FATE—OF THE UNIVERSE

The universe is expanding. Will it forever? Or will it someday reverse itself, start collapsing, and annihilate itself in a "big crunch"? The answer lies in a number astronomers appropriately designate by the last letter of the Greek alphabet, omega.

Omega measures the universe's mass density: the greater the mass density, the greater is omega. Mass exerts gravitational attraction, so it slows the universe's expansion. If the mass density is sufficiently great, it will actually stop the expansion and cause the universe to collapse. On the other hand, if the mass density is low, gravity will only slow the expansion but never stop it, and the universe will expand forever.

The dividing line between these two very different fates is represented by an omega of 1. If omega exceeds 1, then the universe is so dense it will eventually collapse; if omega is less than 1, then the universe is so light it will fly apart. By terrestrial standards, a universe at the critical density of omega equal to 1 is still pretty lightweight: on average, its density is less than an ounce of matter spread through a volume the size of Jupiter. That may not sound like much, but the best observations suggest that the universe does not muster even this much material, so it will expand forever.

THE UNIVERSE is expanding. Will it forever?

Like the Hubble constant, omega affects the age of the universe, because it affects the expansion rate. If the universe had no mass—that is, if omega were 0—then the expansion would not slow at all. But the more mass the universe has, the more its expansion has slowed, and so the faster it once expanded, and the younger it must be. For example, if the Hubble constant is 65, and if omega were 0, then the universe would be 15.0 billion years old. But if omega is 0.1—that is, if the universe has 10 percent of the critical density—it would be 13.5 billion years old, and if omega is 1, it would be only 10.0 billion years old. That last number poses a problem, because the globular clusters are older, suggesting that omega is less than 1 and that the universe will expand forever.

Actually measuring omega is even more difficult than measuring the Hubble constant. There are two ways to do it. First, astronomers can try to add up all the matter in the universe; the more there is, the greater omega must be. Second, they can try to measure how much the universe's expansion is slowing; the more it's slowing, the greater omega must be.

Following the first approach, astronomers begin by detecting light from other galaxies. If every Sun's worth of light corresponded to the Sun's amount of mass, the universe would have hardly any matter at all: omega would be around 0.001, far short of 1. However, most stars shine much more faintly than the Sun, and the combined mass of these lesser lights actually outweighs bright stars like the Sun. Furthermore, the dark halos of our Galaxy and others harbor large quantities of dark matter, which emits no light but contributes to omega. Adding in all this dark matter doesn't elevate omega above about 0.1, which is still far short of the critical density. Galaxy clusters have more dark matter, for they would fly apart if their gravity did not restrain the member galaxies, but even these imply that omega is only 0.2 or so—which still means the universe will expand forever.

The second way to measure omega requires determining how much the expansion is decelerating. This is difficult, because the effect is small, and to see it one must look billions of years into the past–that is, billions of light-years away. Recently astronomers have tried to measure the deceleration by determining the relative distances of distant galaxies that have spawned type Ia supernovae. Supernovae are so luminous they can be seen in galaxies billions of light-years away, and type Ia supernovae are those which arise from white dwarf stars that exceed a mass of 1.4 Suns. Since they all explode at the same mass, they should all be about equally luminous, so their apparent brightness should signal their distance from Earth: the fainter they look, the farther they are. Comparing their distances with the redshifts of their host galaxies can reveal how much the universe's expansion has slowed during the last several billion years. Early results from this work indicate little deceleration of the universe's expansion and thus a low omega, well below the critical value of 1.

There is yet another way to ascertain omega, or at least part of omega. This involves the three light elements–hydrogen, helium, and lithium–manufactured minutes after the big bang, because their exact abundances depend on how dense the universe was then.

This approach implies an omega between 0.01 and 0.10, again far below the critical value. However, there is a catch: this number includes only normal, so-called baryonic matter, which participated in the primordial nuclear reactions. If instead some of the universe's mass is nonbaryonic–for example, consisting of exotic subatomic particles–then it would not have affected primordial nucleosynthesis. Thus the true omega, baryonic plus nonbaryonic, could be higher.

Many theoretical cosmologists believe it is much higher–precisely 1. This belief obviously has no observational basis; after all, every observation suggests omega is much lower. Instead, the belief stems from a theory MIT cosmologist Alan Guth invented in 1979, which postulates that the universe expanded extremely rapidly when it was less than a second old. This rapid expansion, called inflation, can explain why the cosmic microwave background is so smooth across the sky. According to inflationary theory, regions of space now far apart were actually in contact with one another prior to inflation and thus had similar properties. The epoch of inflation then pushed these regions far apart, but they retained similar properties, accounting for the smoothness of the cosmic microwave background.

Unfortunately, inflationary theory makes few predictions, so it is hard to confirm or refute, but it does claim that omega is exactly 1. Thus, inflationary cosmologists have tried to reconcile this prediction with the observations that contradict it. For example, inflationary cosmologists may claim that large amounts of dark material exist where there are no galaxies, and that this extra material pushes omega up to 1. But there is no evidence of this. Furthermore, an omega of 1 implies a universe younger than the oldest stars. If omega is 1, and the Hubble constant 65, the universe is only 10 billion years old–younger than the Milky Way's globular clusters.

FACING PAGE: Galaxy clusters harbor huge amounts of dark matter; this one's gravity distorts the image of a more distant galaxy into blue arcs.

LAMBDA: A RUNAWAY UNIVERSE?

The Hubble constant, which expresses the universe's expansion rate, and omega, which slows it, are the two main cosmological parameters. Many scientists hope they are the *only* cosmological parameters, for few like the third one, which Albert Einstein introduced in 1917. The year before, he had published his great theory of gravity, general relativity, but when he applied the theory to the universe, he found that the weight of the universe caused it to collapse. Einstein therefore postulated that empty space exerts a repulsive force that holds the universe up. This repulsive force is the cosmological constant, or lambda. After astronomers discovered that the universe is expanding, Einstein threw out the cosmological constant, calling it the biggest blunder of his life.

Ever since, the cosmological constant has lived in infamy, a fudge factor concocted merely to make theory agree with observation. But if lambda exists, it opposes omega. Omega represents attraction and slows the universe's expansion. But lambda represents repulsion and accelerates the expansion. And the more the universe expands, the stronger the cosmological constant gets, because a larger universe has more space, from which lambda derives its power. Thus, a lambda-dominated universe does not merely expand; it expands at an ever faster clip, becoming a runaway universe.

Lambda can be quantified. It represents an energy that pervades space, and according to Einstein, energy E equals mass m times the speed of light squared, c^2. Thus, cosmologists express lambda as follows. Imagine turning all the energy that the cosmological constant represents into mass; what omega would that mass give? The answer is the numerical value of lambda.

Like omega, lambda affects the age of the universe, but in the opposite way: while omega makes the universe younger, lambda makes it older. That is because if lambda is large, then the universe is now expanding faster than ever; so it expanded more slowly in the past and is therefore older than the Hubble constant and omega imply. Astronomers who find a high Hubble constant and thus a young age for the universe have sometimes invoked the cosmological constant to make the universe old enough to accommodate the globular clusters. Now that astronomers find a lower numerical value for the Hubble constant and younger ages for the globulars, lambda seems unnecessary.

However, it is not dead, for inflationary cosmologists can use it to save their theory in the face of observations that say omega is not 1. It turns out that a more complicated inflation theory predicts not that omega is 1 but that omega *plus* lambda is 1. For example, if omega is only 0.1, then a lambda of 0.9 would not contradict inflationary cosmology. Thus, the cosmological constant reasserts its familiar role as cosmological fudge factor.

Unfortunately, actually measuring the cosmological constant is extraordinarily difficult. One approach is to observe distant quasars and see whether their light is split by the gravity of galaxies that happen to reside between them and Earth. This so-called gravitational lensing should be common if lambda is large, because then quasars have accelerated away from us, and more space exists between them and Earth, so the chance that an intervening galaxy lenses their light increases. The first gravitationally lensed quasar was discovered in 1979, near the Big Dipper, but since then few others have turned up. This implies a low or zero lambda. Thus, if omega is around 0.1, then the original prediction of inflation, that omega equals 1, is not borne out; nor is the more complicated prediction, that omega plus lambda equals 1.

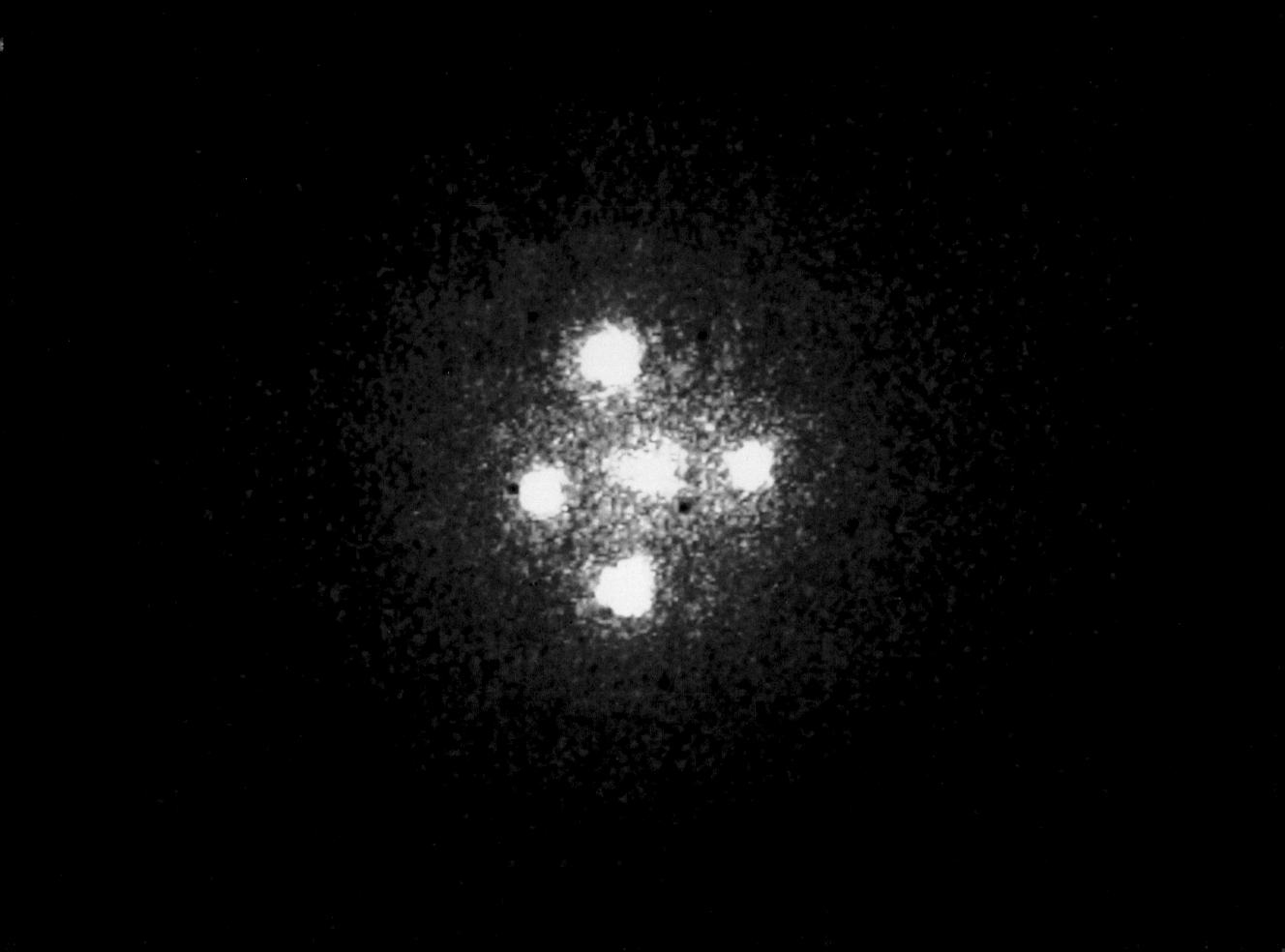

THE UNIVERSE THAT DARED TO DREAM

Whatever the precise numerical values of the three cosmological parameters, the universe has certainly led an interesting existence. It began 10 to 15 billion years ago, when the big bang created the space into which it unleashed a fury of light and matter. Within minutes, the primordial nuclear inferno forged the three lightest elements, hydrogen, helium, and lithium. As the universe expanded, it cooled. The intense light from its birth was trapped by its own electrons, but after the universe cooled enough to allow the electrons to join with protons, the light flooded out, light that astronomers now see at redshift 1,100, the cosmic microwave background. As the universe expanded further, it cooled still more, and the cosmos grew dark, because all its radiation was redshifted to wavelengths longer than those visible. But darkness prevailed only so long. Regions of space that possessed more matter than others collapsed under their weight, giving birth to the first galaxies, and the cosmos began to witness the beauty of starlight, which impinged on the dark. Newborn galaxies sometimes collided and triggered quasars, which proliferated at redshifts 2 and 3. Because of these collisions, some spiral galaxies trans-

formed themselves into ellipticals. The Milky Way, though, survived this traumatic period as a delicate spiral, and so did its large neighbor, the Andromeda Galaxy. Perhaps astronomers on the other side of the observable universe have photographed our Galaxy's birth; perhaps they are beaming these photographs to Earth now, to show us what our Galaxy looked like when it was young. Here, 4.6 billion years ago, at a redshift of 0.5, there arose a beautiful yellow star with nine planets, the third of which became warm and wet and alive.

Life is the universe's most astonishing achievement. The stars shine brightly, planets whirl around them, and both gather into galaxies; but life outdoes them all. Prior to its development, the universe followed the rigid rules of physics and chemistry, which dictated which nuclei could fuse into which others and which molecules could form from which atoms. But with the emergence of life, especially animal life, part of the material world began to think. An animal can choose to run or rest; a human being can choose to read or sleep. Life enriches the universe and infuses it with a spirit and independence it would otherwise lack.

Yet it all came very close to not even happening. Life may be partially spiritual, but its material existence depends on astronomy, physics, and chemistry. For example, the early universe contained nearly as much antimatter as matter, and most of the former annihilated most of the latter; but if the antimatter had exactly equaled the matter, the universe would carry none of either, and life would not exist. Carbon, central to life on Earth, arises from a nuclear reaction that fuses helium. This reaction in turn depends on a property of the carbon atom, called a resonance, which happens to occur at almost exactly the same energy as the centers of the stars which produce it. Without this coincidence, little carbon would exist, and the cosmos would be dead.

FACING PAGE: The gravity of a galaxy (central object) splits light from a quasar behind it into four images. The larger the cosmological constant,

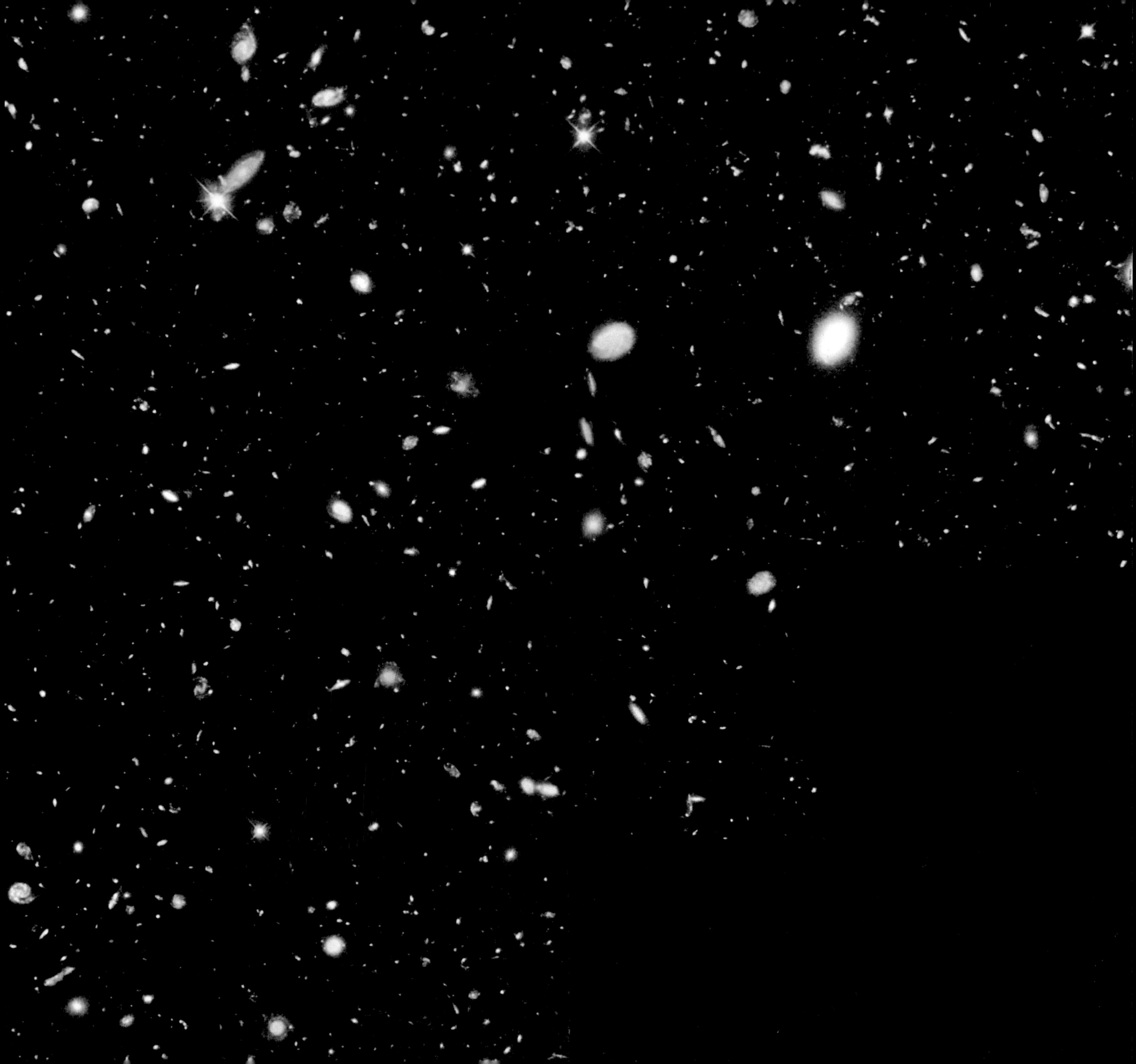

Life's existence also depends on the numbers that govern basic physics. These so-called physical constants include the speed of light, c; the charge of an electron, e; the strength of gravity, G; and the scale of quantum effects, which is set by the Planck constant, h.

No one knows why c, e, G, and h have the particular values they do—why c is large, G and h small—but if they did not, life would probably not exist. For example, because G is small, gravity is weak. If G were instead large, the Sun would have to burn its fuel much faster in order to support its weight, so it would have burned out long ago. If G were smaller and gravity weaker, then stars might not have been able to form, since they arose via gravitational collapse. Similar arguments apply to c, e, and h. Also crucial are the astronomical constants, such as the Hubble constant and omega. If the Hubble constant were too high, the universe would have flown apart so fast that matter would not have conglomerated into galaxies; and if omega were too high, the universe would have collapsed long ago. Thus, we owe our existence to a vast cascade of cosmic coincidences.

EARTH as it is
the heavens

Why? Why does the cosmos have exactly the right physical and astronomical constants for life? Did a creator deliberately choose these numbers so that the cosmos would spawn life? Or is our universe one of countless universes, each with different physical and astronomical constants? Occam's razor—the principle that simpler hypotheses tend to be right—favors the former explanation, which suggests the existence of God, but the latter explanation hardly rules out such a being. If God did indeed create the universe, it must reflect God's own nature. God must be a creative being, for the universe itself is, sculpting what was originally amorphous matter and light into stars and galaxies and planets. And God must love diversity, for the cosmos abounds with it: stars red and yellow and blue, young and old, huge and tiny; galaxies titanic and dwarf, quiet and restless, elliptical and spiral; planets gaseous and rocky, big and little, living and dead. On Earth as it is in the heavens: terrestrial life—and human beings—boast the same diversity, with people Asian, white, and black; gay and straight; young and old. Indeed, humanity is a microcosm of the vibrant color that pervades the entire cosmos.

The universe is a remarkable place—remarkable not only for its grandeur but also for its varied offspring. The planets will continue to circle the Sun, the Sun to orbit the Galaxy, and most galaxies to recede from ours; but unlike them, we choose our own destiny.

A magnificent universe it is.

THE SUN'S PLANETS

| Planet | Mean Distance from the Sun | | Year | Day | Axial Tilt (degrees) |
	AU*	Miles	Kilometers			
Mercury	0.3871	36,000,000	57,900,000	88 d	58.65 d	0
Venus	0.7233	67,200,000	108,200,000	225 d	243.0 d	177.4
Earth	1.0000	92,960,000	149,600,000	365¼ d	23 h 56 m	23.4
Mars	1.5237	141,600,000	227,900,000	687 d	24 h 37 m	25.2
Jupiter	5.203	483,600,000	778,300,000	11.9 y	9 h 55 m	3.1
Saturn	9.54	886,700,000	1,427,000,000	29.5 y	10 h 39 m	26.7
Uranus	19.2	1,780,000,000	2,870,000,000	84.0 y	17 h 14 m	97.8
Neptune	30.1	2,790,000,000	4,500,000,000	165 y	16 h 7 m	28.3
Pluto	39.5	3,670,000,000	5,910,000,000	248 y	6.39 d	122.5

Orbital Eccentricity**	Orbital Inclination***	Mass (Earth=1)	Equatorial Diameter Miles	Equatorial Diameter Kilometers	Mean Density (water=1)	Temperature (Fahrenheit)	Number of Moons
0.206	7.0	0.055	3,032	4,879	5.43	+250	0
0.007	3.4	0.815	7,521	12,104	5.24	+860	0
0.017	0.0	1.000	7,926	12,756	5.52	+60	1
0.093	1.8	0.107	4,222	6,794	3.93	−67	2
0.048	1.3	317.83	88,846	142,984	1.33	−236	16
0.054	2.5	95.16	74,898	120,536	0.69	−289	18
0.047	0.8	14.54	31,763	51,118	1.27	−353	18
0.009	1.8	17.15	30,775	49,528	1.64	−353	8
0.25	17.1	0.002	1,430	2,300	2.0	−400?	1

*One AU—astronomical unit—is the mean distance from Sun to Earth.

**Eccentricity measures orbital shape: a circular orbit has eccentricity 0.00 and an elliptical orbit a greater eccentricity; the most elongated bound orbit has an eccentricity just under 1.00.

***Incl nation measures the tilt, in degrees, between the plane of a planet's orbit and the plane of the Earth's orbit.

MOONS

Planet	Moons Number	Name	Discovery Year	Mean Distance from Planet Miles	Kilometers	Orbital Period	Mean Diameter Miles	Kilometers
EARTH	1	Moon	—	238,900	384,400	27.322 d	2,159	3,475
MARS	2	Phobos	1877	5,827	9,378	7 h 39 m	14	22
		Deimos	1877	14,577	23,459	1.262 d	8	12
JUPITER	16	Metis	1979	79,500	128,000	7 h 4 m	27	43
		Adrastea	1979	80,200	129,000	7 h 9 m	10	16
		Amalthea	1892	112,000	181,000	11 h 57 m	104	167
		Thebe	1979	138,000	222,000	16 h 11 m	61	99
		Io	1610	262,000	422,000	1.769 d	2,259	3,636
		Europa	1610	417,000	671,000	3.551 d	1,940	3,121
		Ganymede	1610	665,000	1,070,000	7.155 d	3,274	5,268
		Callisto	1610	1,170,000	1,883,000	16.689 d	2,993	4,817
		Leda	1974	6,893,000	11,094,000	238.72 d	6	10
		Himalia	1904	7,133,000	11,480,000	250.57 d	106	170
		Lysithea	1938	7,282,000	11,720,000	259.22 d	15	24
		Elara	1905	7,293,000	11,737,000	259.65 d	50	80
		Ananke*	1951	13,200,000	21,200,000	631 d	12	20
		Carme*	1938	14,000,000	22,600,000	692 d	19	30
		Pasiphae*	1908	14,600,000	23,500,000	735 d	22	36
		Sinope*	1914	14,700,000	23,700,000	758 d	17	28
SATURN	18	Pan	1990	83,005	133,583	13 h 48 m	12	20
		Atlas	1980	85,540	137,670	14 h 27 m	20	32
		Prometheus	1980	86,590	139,353	14 h 43 m	62	100
		Pandora	1980	88,050	141,700	15 h 5 m	52	84
		Epimetheus	1966	94,089	151,422	16 h 40 m	74	119
		Janus	1966	94,120	151,472	16 h 40 m	110	178
		Mimas	1789	115,280	185,520	22 h 37 m	247	397
		Enceladus	1789	147,900	238,020	1.370 d	310	499
		Tethys	1684	183,090	294,660	1.888 d	658	1,060
		Telesto	1980	183,090	294,660	1.888 d	14	22
		Calypso	1980	183,090	294,660	1.888 d	12	19
		Dione	1684	234,500	377,400	2.737 d	696	1,120
		Helene	1980	234,500	377,400	2.737 d	20	32

TABLE TWO

Planet	Moons Number	Name	Discovery Year	Mean Distance from Planet (Miles)	Mean Distance from Planet (Kilometers)	Orbital Period	Mean Diameter (Miles)	Mean Diameter (Kilometers)
SATURN		Rhea	1672	327,490	527,040	4.518 d	949	1,528
		Titan	1655	759,210	1,221,830	15.945 d	3,200	5,150
		Hyperion	1848	920,310	1,481,100	21.277 d	176	283
		Iapetus	1671	2,212,900	3,561,300	79.330 d	892	1,436
		Phoebe*	1899	8,048,000	12,952,000	550.48 d	137	220
URANUS	18	Cordelia	1986	30,910	49,750	8 h 2 m	16	26
		Ophelia	1986	33,410	53,760	9 h 2 m	19	30
		Bianca	1986	36,760	59,170	10 h 26 m	26	42
		Cressida	1986	38,380	61,770	11 h 8 m	39	62
		Desdemona	1986	38,930	62,660	11 h 22 m	34	54
		Juliet	1986	39,990	64,360	11 h 50 m	52	84
		Portia	1986	41,070	66,100	12 h 19 m	67	108
		Rosalind	1986	43,450	69,930	13 h 24 m	34	54
		Belinda	1986	46,760	75,260	14 h 58 m	41	66
		———	1999	47,480	76,420	15 h 18 m	25	40
		Puck	1985	53,440	86,000	18 h 17 m	96	154
		Miranda	1948	80,400	129,390	1.413 d	293	472
		Ariel	1851	118,690	191,020	2.520 d	719	1,158
		Umbriel	1851	165,470	266,300	4.144 d	727	1,169
		Titania	1787	270,860	435,910	8.706 d	980	1,578
		Oberon	1787	362,580	583,520	13.463 d	946	1,523
		Caliban*	1997	4,000,000	7,000,000	600 d	40	60
		Sycorax*	1997	8,000,000	12,000,000	1300 d	80	120
NEPTUNE	8	Naiad	1989	29,970	48,230	7 h 4 m	36	58
		Thalassa	1989	31,110	50,070	7 h 29 m	50	80
		Despina	1989	32,640	52,530	8 h 2 m	92	148
		Galatea	1989	38,490	61,950	10 h 17 m	98	158
		Larissa	1989	45,700	73,550	13 h 19 m	119	192
		Proteus	1989	73,100	117,650	1.122 d	258	416
		Triton*	1846	220,440	354,760	5.877 d	1,681	2,705
		Nereid	1949	3,425,900	5,513,400	360.14 d	210	340
PLUTO	1	Charon	1978	12,200	19,600	6.387 d	750	1,200

*Retrograde: moon orbits planet opposite from the direction of planet's spin.

| Rank | Star | Constellation | Distance (light-years) | Spectral Type* | Magnitudes** | | Visible Light Output (Sun=1) |
					Apparent Visual	Absolute Visual	
0	Sun	—	0	G2V	−26.74	+4.83	1.0
1	Sirius	Canis Major	8.6	A1 V	−1.44	+1.45	22.4
2	Canopus	Carina	310	F0 Ib	−0.62	−5.5	14,000
3	Alpha Centauri	Centaurus	4.4	G2V/K1 V	−0.28	+4.34	1.57
4	Arcturus	Boötes	37	K2 III	−0.05	−0.31	113
5	Vega	Lyra	25	A0V	+0.03	+0.58	50
6	Capella	Auriga	42	G8 III/G0 III	0.08	+0.20	71
7	Rigel	Orion	800	B8 Ia	0.18	−6.8	44,000
8	Procyon	Canis Minor	11.4	F5 IV–V	0.40	+2.68	7.2
9	Achernar	Eridanus	144	B3V	0.45	−2.8	1,100
9	Betelgeuse	Orion	430	M2 Ib	0.45	−5.1	9,700
11	Beta Centauri	Centaurus	530	B1 III	0.61	−5.4	13,000
12	Altair	Aquila	16.8	A7V	0.76	+2.20	11
13	Alpha Crucis	Crux	320	B0.5 IV/B1 V	0.77	−3.6	2,300
14	Aldebaran	Taurus	65	K5 III	0.87	−0.63	150
15	Spica	Virgo	260	B1 IV	0.98	−3.4	2,000
16	Antares	Scorpius	600	M1 Ib	1.06	−5.3	11,000
17	Pollux	Gemini	34	K0 III	1.16	+1.09	31
18	Fomalhaut	Piscis Austrinus	25	A3 V	1.17	+1.74	17
19	Deneb	Cygnus	1,800	A2 Ia	1.25	−7.6	94,000
19	Beta Crucis	Crux	350	B0.5 III	1.25	−3.9	3,200
21	Regulus	Leo	77	B7V	1.36	−0.52	140
22	Adhara	Canis Major	430	B2 II	1.50	−4.1	3,700
23	Castor	Gemini	52	A1 V/A2V	1.58	+1.0	34
24	Gamma Crucis	Crux	88	M3.5 III	1.59	−0.56	140
25	Shaula	Scorpius	700	B2 IV/B	1.62	−4.3	4,500

*The roman numeral following the spectral type gives the luminosity class:
I supergiant   II bright giant   III giant   IV subgiant   V main-sequence star.
If a second star in the system contributes significant light, spectral types and luminosity classes of both stars appear.

**The brighter a star, the smaller the magnitude; negative magnitudes are the brightest. Apparent magnitude is how
bright a star looks; absolute magnitude reveals how bright it really is—it is the apparent magnitude the star would
have if it were 32.6 light-years from Earth. For multiple stars, the apparent magnitude in Table 3 is the combined
magnitudes of all stars in the system, whereas the absolute magnitude refers only to the system's brightest star.

THE NEAREST STARS

Star	Constellation	Distance (light-years)	Spectral Type*	Apparent Visual	Absolute Visual	Visible Light Output (Sun=1)
Sun	—	0.00	G2V	−26.74	+4.83	1.00
Alpha Centauri A	Centaurus	4.40	G2V	−0.01	+4.34	1.57
Alpha Centauri B	Centaurus	4.40	K1V	+1.35	5.70	0.45
Alpha Centauri C	Centaurus	4.24	M5.5V	+11.01	15.44	0.000057
Barnard's Star	Ophiuchus	5.94	M4V	9.54	13.24	0.00043
Wolf 359	Leo	7.8	M6V	13.45	16.56	0.0000020
Lalande 21185	Ursa Major	8.3	M2V	7.49	10.46	0.0056
Sirius A	Canis Major	8.6	A1 V	−1.44	+1.45	22.4
Sirius B	Canis Major	8.6	WD	+8.4	11.3	0.0026
Luyten 726-8 A	Cetus	8.7	M5.5V	12.54	15.40	0.00006
Luyten 726-8 B	Cetus	8.7	M6V	12.99	15.85	0.0000-4
Ross 154	Sagittarius	9.7	M3.5V	10.37	13.00	0.00054
Ross 248	Andromeda	10.3	M5.5V	12.29	14.79	0.00010
Epsilon Eridani	Eridanus	10.5	K2V	3.72	6.18	0.29
Lacaille 9352	Piscis Austrinus	10.7	M1.5V	7.35	9.76	0.011
Ross 128	Virgo	10.9	M4V	11.12	13.50	0.00034
Luyten 789-6 A	Aquarius	11.3	M5V	12.85**	15.16**	0.00007**
Luyten 789-6 B		11.3				
Luyten 789-6 C		11.3				
61 Cygni A	Cygnus	11.4	K5V	5.20	7.48	0.087
61 Cygni B	Cygnus	11.4	K7V	6.05	8.33	0.040
Procyon A	Canis Minor	11.4	F5 IV–V	0.40	2.68	7.2
Procyon B	Canis Minor	11.4	WD	10.9	13.2	0.0004
Struve 2398 A	Draco	11.6	M3V	8.94	11.19	0.0029
Struve 2398 B	Draco	11.6	M3.5V	9.70	11.95	0.0014
Groombridge 34 A	Andromeda	11.6	M1.5V	8.09	10.33	0.0063
Groombridge 34 B	Andromeda	11.6	M3.5V	11.07	13.31	0.00041
Giclas 51-15	Cancer	11.8	M6.5V	14.79	16.99	0.000014
Epsilon Indi	Indus	11.8	K4.5V	4.69	6.89	0.15
Tau Ceti	Cetus	11.9	G8V	3.49	5.68	0.46

* See notes to Table 3.

** Combined output of Luyten 789-6 A and C.

LOCAL GROUP GALAXIES

Galaxy	Alternate Name	Constellation	Type*	Distance from Sun (light-years)	Absolute Visual Magnitude	Output of Visible Light**
MILKY WAY'S EMPIRE						
Milky Way	—	—	Sbc	—	−20.6	1.0
Sagittarius***	—	Sagittarius	dSph	80,000	−14.0	0.002
Large Magellanic Cloud	—	Dorado	Irr	160,000	−18.1	0.1
Small Magellanic Cloud	—	Tucana	Irr	190,000	−16.2	0.02
Ursa Minor	—	Ursa Minor	dSph	220,000	−8.5	0.00001
Draco	—	Draco	dSph	250,000	−8.4	0.00001
Sculptor	—	Sculptor	dSph	260,000	−10.9	0.0001
Sextans	—	Sextans	dSph	280,000	−9.2	0.00003
Carina	—	Carina	dSph	300,000	−8.8	0.00002
Fornax	—	Fornax	dSph	420,000	−13.2	0.001
Leo II	—	Leo	dSph	690,000	−9.6	0.00004
Leo I	—	Leo	dSph	740,000	−11.8	0.0003
ANDROMEDA'S EMPIRE						
Andromeda	M31	Andromeda	Sb	2.4 million	−21.1	1.6
M33	—	Triangulum	Sc	2.6 million	−18.9	0.2
M32	—	Andromeda	E2	2.4 million	−16.4	0.02
NGC 205	—	Andromeda	E5	2.4 million	−16.3	0.02
NGC 185	—	Cassiopeia	E3	2.0 million	−15.2	0.007
NGC 147	—	Cassiopeia	E5	2.2 million	−15.1	0.006
Andromeda I	—	Andromeda	dSph	2.6 million	−12.0	0.0004
Andromeda II	—	Pisces	dSph	2.3 million	−11.7	0.0003
Andromeda III	—	Andromeda	dSph	2.4 million	−10.3	0.00008
Pisces	LGS 3	Pisces	dSph	2.6 million	−10.3	0.00008
Andromeda V	—	Andromeda	dSph	2.6 million		
Andromeda VI	Pegasus II	Pegasus	dSph	2.5 million		
Andromeda VII	Cassiopeia	Cassiopeia	dSph	2.3 million		

Hertzsprung-Russell (H-R) Diagram   A plot of stellar luminosity against stellar temperature, color, or spectral type. The H-R diagram places stars into three categories: main-sequence stars, like the Sun; giants and supergiants, like Arcturus and Antares; and white dwarfs, like Sirius B.

Hubble Constant   The present expansion rate of the universe. The higher the Hubble constant, the faster the universe is expanding and the younger it is.

Interstellar Medium   The space between the stars, and the home of nebulae.

Irregular Galaxy   An amorphous galaxy, like the Large Magellanic Cloud, that bears large amounts of gas, star formation, and young stars.

Kuiper Belt   The zone of icy bodies just beyond Neptune's orbit that supplies most short-period comets to the solar system.

Lambda (λ)   The numerical value of the cosmological constant, the repulsive force of empty space. If the cosmological constant exists, it accelerates the universe's expansion.

Light-Year   The distance light travels in one year: 5.88 trillion miles, or 9.46 trillion kilometers. The closest star to the Sun is just over 4 light-years away.

Local Group   The group of about three dozen nearby galaxies that includes the Milky Way, Andromeda, M33, and the Magellanic Clouds.

Luminosity   The true, or intrinsic, power of a star—as opposed to how bright it looks, which depends on both its power and its distance.

Magnitude   A measure of how bright a star is (absolute magnitude) or looks (apparent magnitude). The brighter the star, the *smaller* the magnitude, so a first-magnitude star outdoes a second-magnitude star.

Main-Sequence Star   A star, like the Sun, that generates energy by burning hydrogen into helium at its center.

Meteor   The streak of light produced when a particle in space—a meteoroid—burns up in a planet's atmosphere. If the particle reaches the ground, the surviving object is called a meteorite.

Microwave Background   The big bang's afterglow, preserving a record of the universe as it was just a few hundred thousand years after its birth.

Milky Way   Our Galaxy—the home of the Sun, the Earth, and every individual star the naked eye can see.

Mira   1. The name of a pulsating red giant star in Cetus. 2. Any red giant that pulsates.

Moon   A natural body orbiting a planet. The Earth has one moon; most planets have more.

Multiple Star   Two or more stars that revolve around one another.

Nebula   A cloud of interstellar gas and dust. Some nebulae, such as H II regions, spawn new stars, while some, such as supernova remnants and planetary nebulae, are the remains of dead or dying stars.

Neutron Star   A collapsed star, ten miles wide, that outweighs the Sun and consists mostly of neutrons. Neutron stars are sometimes pulsars.

Nova   An explosion in a binary star that causes it to brighten dramatically but temporarily.

Nucleosynthesis   The transformation of old elements into new—such as a red giant's conversion of helium into carbon and oxygen.

Olbers' Paradox   An infinitely old, infinitely extended universe filled with stars and galaxies should look bright at night; that it does not is called Olbers' paradox. The night is dark in part because the universe had a beginning.

Omega (Ω)   The density of the universe, expressed as a fraction of the density that would cause it to someday collapse. If omega is less than 1, the universe has a low density and will expand forever; if omega is greater than 1, the universe is so dense it will someday collapse.

Oort Cloud   The cometary zone that surrounds the solar system, lies far beyond the Edgeworth-Kuiper belt, and supplies long-period comets.

Open Cluster   A loose gathering of a few hundred stars. The most famous open clusters are the Pleiades and the Hyades.

Parallax   The small shift in a star's apparent position that results as the Earth circles the Sun. Parallax reveals stellar distances, because the greater the parallax, the closer is the star.

Planet   An object that formed in a disk around a star. Unlike stars, planets emit no light of their own; they merely reflect the light of the star(s) they orbit.

Planetary Nebula   A red giant star's cast-off atmosphere, set aglow by the star's newly exposed hot core. The most famous planetary nebula is the Ring Nebula in Lyra.

Population   A galaxy-wide assemblage of stars that share similar ages, locations, velocities, and chemical abundances. The Milky Way has four stellar populations: the thin disk, the thick disk, the stellar halo, and the bulge.

Primordial Nucleosynthesis   The creation of elements during the first few minutes of the universe's life, which gave birth to the three lightest elements: hydrogen, helium, and lithium.

Pulsar   A fast-spinning neutron star that emits radio waves.

Quasar   A galaxy center that can radiate a trillion times more light than the Sun. Quasars proliferated long ago, when the universe was young.

Radio Galaxy A galaxy that emits large amounts of radio waves. The nearest radio galaxy is Centaurus A.

Reddening The absorption and scattering of blue light by interstellar dust, making stars look redder than they really are.

Red Dwarf A small, faint, cool star of spectral type M. Red dwarfs outnumber all other main-sequence stars put together, yet they are so faint that none is visible to the unaided eye.

Red Giant A large, bright, cool star that has evolved off the main sequence and no longer burns hydrogen in its core.

Redshift A spectral shift to longer wavelengths, caused by (a) an object's movement away from us (Doppler shift); (b) the expansion of space between it and us (cosmological redshift); or (c) gravity (gravitational redshift).

Ring Galaxy A rare galaxy that forms when a small galaxy smashes through a larger one's center and creates a ring of star formation that propagates outward from the large galaxy's center.

Satellite A small object orbiting a larger one—such as a moon circling a planet or a satellite galaxy orbiting a giant galaxy.

Solar System The planets, moons, asteroids, and comets that a star rules.

Spectral Type An indicator of a star's temperature and color. There are seven main spectral types, which from hot and blue to cool and red are O (blue), B (also blue), A (white), F (yellow-white), G (yellow), K (orange), and M (red).

Spiral Galaxy A galaxy, like the Milky Way and Andromeda, that resembles a pinwheel and hosts stars with a variety of ages.

Star A self-luminous object, like the Sun.

Starburst Galaxy A galaxy giving birth to many new stars.

Star Cluster A gathering of stars. There are two types: open clusters and globular clusters.

Stellar Halo The old stellar population that rises above and below a spiral galaxy's disk and in the Milky Way lies primarily interior to the Sun's orbit around the Galaxy.

Stellar Population A galaxy-wide assemblage of stars that share similar ages, locations, velocities, and chemical abundances. The Milky Way has four stellar populations: the thin disk, the thick disk, the stellar halo, and the bulge.

Supercluster A vast assemblage of galaxies, consisting of at least one galaxy cluster and numerous galaxy groups.

Supergiant A large, extremely luminous star that no longer fuses hydrogen at its center and will soon explode. Examples include Rigel, Deneb, Antares, and Betelgeuse.

Supernova A titanic explosion that tears a star apart. Type Ia supernovae arise from exploding white dwarfs; type Ib, Ic, and II supernovae arise from exploding high-mass stars.

Supernova Remnant The remains of an exploded star. The most famous supernova remnant is the Crab Nebula in Taurus.

Surface Brightness A measure of how spread out a galaxy's light is. All famous galaxies, like the Milky Way and Andromeda, have high surface brightnesses; low-surface-brightness galaxies are much harder to see.

Thick Disk The old stellar population whose stars reside in a disk but journey farther from the Galaxy's plane than thin-disk stars do.

Thin Disk The dominant stellar population near the Sun. Thin-disk stars have various ages, fairly circular orbits around the Galaxy, and high abundances of elements like oxygen and iron.

White Dwarf The small, faint, dying cinder of a Sunlike star, so dense that a spoonful would weigh tons.

FURTHER READING

SUN

Kippenhahn, Rudolf, 1994. *Discovering the Secrets of the Sun* (Chichester: Wiley).

MERCURY

Strom, Robert G., 1987. *Mercury* (Washington, D.C.: Smithsonian Institution Press).

VENUS

Grinspoon, David Harry, 1997. *Venus Revealed* (Reading, Massachusetts: Addison-Wesley).

EARTH

Hartmann, William K., and Miller, Ron, 1991. *The History of Earth* (New York: Workman).

MOON

Chaikin, Andrew, 1994. *A Man on the Moon* (New York: Viking).

Wilhelms, Don E., 1993. *To a Rocky Moon* (Tucson: University of Arizona Press).

MARS

Raeburn, Paul 1998. *Mars* (Washington, D.C.: National Geographic Society).

ASTEROIDS

Kowal, Charles T., 1996. *Asteroids*, Second Edition (Chichester: Wiley).

JUPITER

Morrison, David, and Samz, Jane, 1980. *Voyage to Jupiter* (Washington, D.C.: NASA).

SATURN

Morrison, David, 1982. *Voyages to Saturn* (Washington, D.C.: NASA).

URANUS

Hunt, Garry, and Moore, Patrick, 1989. *Atlas of Uranus* (Cambridge: Cambridge University Press).

NEPTUNE

Grosser, Morton, 1962. *The Discovery of Neptune* (Cambridge: Harvard University Press).

Hunt, Garry E., and Moore, Patrick, 1994. *Atlas of Neptune* (Cambridge: Cambridge University Press).

PLUTO

Stern, Alan, and Mitton, Jacqueline, 1998. *Pluto and Charon* (New York: Wiley).

Tombaugh, Clyde W., and Moore, Patrick, 1980. *Out of the Darkness* (Harrisburg, Pennsylvania: Stackpole).

COMETS

Schaaf, Fred, 1997. *Comet of the Century* (New York: Copernicus).

INTERSTELLAR MEDIUM

Kaler, James B., 1997. *Cosmic Clouds* (New York: Freeman).

Wynn-Williams, Gareth, 1992. *The Fullness of Space* (Cambridge: Cambridge University Press).

STAR BIRTH

Cohen, Martin, 1988. *In Darkness Born* (Cambridge: Cambridge University Press).

STELLAR EVOLUTION

Kaler, James B., 1992. *Stars* (New York: Freeman).

BLACK HOLES

Thorne, Kip S., 1994. *Black Holes and Time Warps* (New York: Norton).

EXTRASOLAR PLANETS

Croswell, Ken, 1997. *Planet Quest* (New York: The Free Press).

EXTRATERRESTRIAL LIFE

de Duve, Christian, 1995. *Vital Dust* (New York: Basic).

Dick, Steven J., 1996. *The Biological Universe* (Cambridge: Cambridge University Press).

Zuckerman, Ben, and Hart, Michael H. (editors), 1995. *Extraterrestrials: Where Are They?*, Second Edition (Cambridge: Cambridge University Press).

MILKY WAY

Croswell, Ken, 1995. *The Alchemy of the Heavens* (New York: Doubleday/Anchor).

Henbest, Nigel, and Couper, Heather, 1994. *The Guide to the Galaxy* (Cambridge: Cambridge University Press).

GALAXIES

Hodge, Paul W., 1986. *Galaxies* (Cambridge: Harvard University Press).

Sandage, Allan, and Bedke, John, 1994. *The Carnegie Atlas of Galaxies* (Washington, D.C.: Carnegie Institution of Washington).

van den Bergh, Sidney, 1998. *Galaxy Morphology and Classification* (Cambridge: Cambridge University Press).

OLBERS' PARADOX

Harrison, Edward, 1987. *Darkness at Night* (Cambridge: Harvard University Press).

BIG BANG VERSUS STEADY STATE

Kragh, Helge, 1996. *Cosmology and Controversy* (Princeton: Princeton University Press).

COSMIC MICROWAVE BACKGROUND

Chown, Marcus, 1996. *Afterglow of Creation* (Sausalito, California: University Science Books).

Mather, John C., and Boslough, John, 1996. *The Very First Light* (New York: Basic).

Smoot, George, and Davidson, Keay, 1993. *Wrinkles in Time* (New York: Morrow).

GENERAL COSMOLOGY

Lightman, Alan, and Brawer, Roberta, 1990. *Origins* (Cambridge: Harvard University Press).

Overbye, Dennis, 1991. *Lonely Hearts of the Cosmos* (New York: HarperCollins).

Rees, Martin, 1995. *Perspectives in Astrophysical Cosmology* (Cambridge: Cambridge University Press).

As indicated below, most photographs are copyrighted; these may not be reproduced without the permission of the copyright holders. For more information on the copyright holders, see http://www.ccnet.com/~galaxy.

Page	Object	Credit
THE PLANETS		
5	Sun	© Association of Universities for Research in Astronomy (AURA), National Optical Astronomy Observatories (NOAO)/National Solar Observatory, Sacramento Peak/National Science Foundation (NSF).
7	Mercury	NASA/Mariner 10. Courtesy Robert G. Strom.
9	Venus	NASA/Pioneer Venus. Courtesy National Space Science Data Center (NSSDC).
11	Earth	NASA/Apollo 17.
12	Volcano	© Tony Stone Images.
13	Forest	© Ric Ergenbright/Tony Stone Images.
14	Aurora	© Jay Brausch.
16	Moon	© UCO/Lick Observatory photo.
18	Mars globe	NASA/Viking/U.S. Geological Survey. Courtesy NSSDC.
20	Mars surface	NASA/Mars Pathfinder. Courtesy NSSDC.
22	Phobos	NASA/Viking. Courtesy NSSDC.
23	Ida	NASA/Galileo. Courtesy NSSDC.
25	Jupiter	NASA/Voyager 1. Courtesy NSSDC.
27	Io	NASA/Galileo. Courtesy NSSDC.
28	Europa	NASA/Galileo. Courtesy NSSDC.
30	Yellow Saturn	NASA/Voyager 2. Courtesy NSSDC.
32	White Saturn	NASA/Voyager 1. Courtesy NSSDC.
33	Titan	NASA/Voyager 2. Courtesy NSSDC.
35	Uranus	NASA/Voyager 2. Courtesy NSSDC.
36	Neptune	NASA/Voyager 2. Courtesy NSSDC.
37	Triton photograph	NASA/Voyager 2. Courtesy NSSDC.
38	Triton painting	© Michael Carroll.
40	Pluto	© AURA. Hubble Space Telescope. R. Albrecht (ESA/ESO Space Telescope European Coordinating Facility) and NASA.
43	Comet Hyakutake	© Gerald Rhemann and Franz Kersche.
44	Comet Hale-Bopp	© Kenneth Lum.
47	Meteor in Orion	© Glendon L. Howell.
THE STARS		
49	California Nebula	© George Greaney.
50	Big and Little Dippers	© Akira Fujii.
52	Horsehead Nebula	© Anglo-Australian Observatory/Royal Observatory Edinburgh. Photograph made from UK Schmidt plates by David Malin.
55	Orion Nebula I	© Mike Sisk.
56	Orion Nebula II	© AURA. Hubble Space Telescope. C. R. O'Dell and S. K. Wong (Rice University), and NASA.
59	Lagoon Nebula	© AURA. NOAO/NSF.
60	Trifid Nebula	© Anglo-Australian Observatory. Photography by David Malin.
61	NGC 6559/IC 1274-75	© Anglo-Australian Observatory/Royal Observatory Edinburgh. Photograph made from UK Schmidt plates by David Malin.

Page	Object	Credit
THE GALAXIES		
109	NGC 2997	© Anglo-Australian Observatory. Photography by David Malin.
111	M87	© Anglo-Australian Observatory. Photography by David Malin.
113	NGC 1232	© European Southern Observatory.
115	NGC 1566	© Anglo-Australian Observatory. Photography by David Malin.
116	NGC 7742	© AURA. Hubble Space Telescope. Courtesy of the Hubble Heritage Team using data collected by C. R. Lynds (NOAO/KPNO) and collaborators, and NASA.
117	NGC 891	© IAC. Photo made from Isaac Newton Telescope plates by David Malin.
118	NGC 2442	© Anglo-Australian Observatory. Photography by David Malin.
120	Barnard's Galaxy	© Anglo-Australian Observatory. Photography by David Malin.
124	Milky Way	© Dennis di Cicco/*Sky and Telescope*.
125	NGC 4565	© Thomas Montemayor. University of Texas McDonald Observatory.
127	Milky Way/Sagittarius	© Peter Ledlie.
128	Milky Way/Orion to Crux	© Akira Fujii.
130	Large Magellanic Cloud	© Anglo-Australian Observatory/Royal Observatory Edinburgh. Photograph made from UK Schmidt plates by David Malin.
131	LMC/Tarantula Nebula	© AURA. Hubble Space Telescope. NASA.
132	Small Magellanic Cloud	© Stefan Binnewies/Bernd Schröter/Peter Riepe/Harald Tomsik.
133	Leo I	© Anglo-Australian Observatory. Photography by David Malin.
134	Andromeda Galaxy	© Tony and Daphne Hallas.
137	M33	© IAC. Photo made from Isaac Newton Telescope plates by David Malin.
138	M33/NGC 604	© AURA. Hubble Space Telescope. H. Yang (University of Illinois) and J. Hester (Arizona State University), and NASA.
140	NGC 253	© Anglo-Australian Observatory. Photography by David Malin.
141	NGC 300	© Anglo-Australian Observatory. Photography by David Malin.
142	Dwingeloo 1	© Dwingeloo Obscured Galaxy Survey team (R. Kraan-Korteweg et al.) and S. Hughes and S. Maddox (Royal Greenwich Observatory).
143	M81	© Tony and Daphne Hallas.
144	M81/M82	© Carl and Chris Weber.
145	Centaurus A I	© Anglo-Australian Observatory. Photography by David Malin.
146	Centaurus A II	© AURA. Hubble Space Telescope. E. J. Schreier (STScI) and NASA.
148	M83	© Anglo-Australian Observatory. Photography by David Malin.
149	M51	© Canada-France-Hawaii Telescope Corporation; photographer, Tom Gregory. Courtesy Richard Talcott, *Astronomy*.
150	M101	© Tony and Daphne Hallas.
151	Virgo Cluster	© Anglo-Australian Observatory/Royal Observatory Edinburgh. Photograph made from UK Schmidt plates by David Malin.
152	Sombrero Galaxy	© Anglo-Australian Observatory. Photography by David Malin.
153	Fornax Cluster	© Anglo-Australian Observatory/Royal Observatory Edinburgh. Photograph made from UK Schmidt plates by David Malin.
154	NGC 1365	© Anglo-Australian Observatory. Photography by David Malin.
158	Antennae Galaxies	© AURA. Hubble Space Telescope. Brad Whitmore (STScI) and NASA.
160	Cartwheel Galaxy	© AURA. Hubble Space Telescope. Kirk Borne (STScI) and NASA.

References to photographs are italicized.